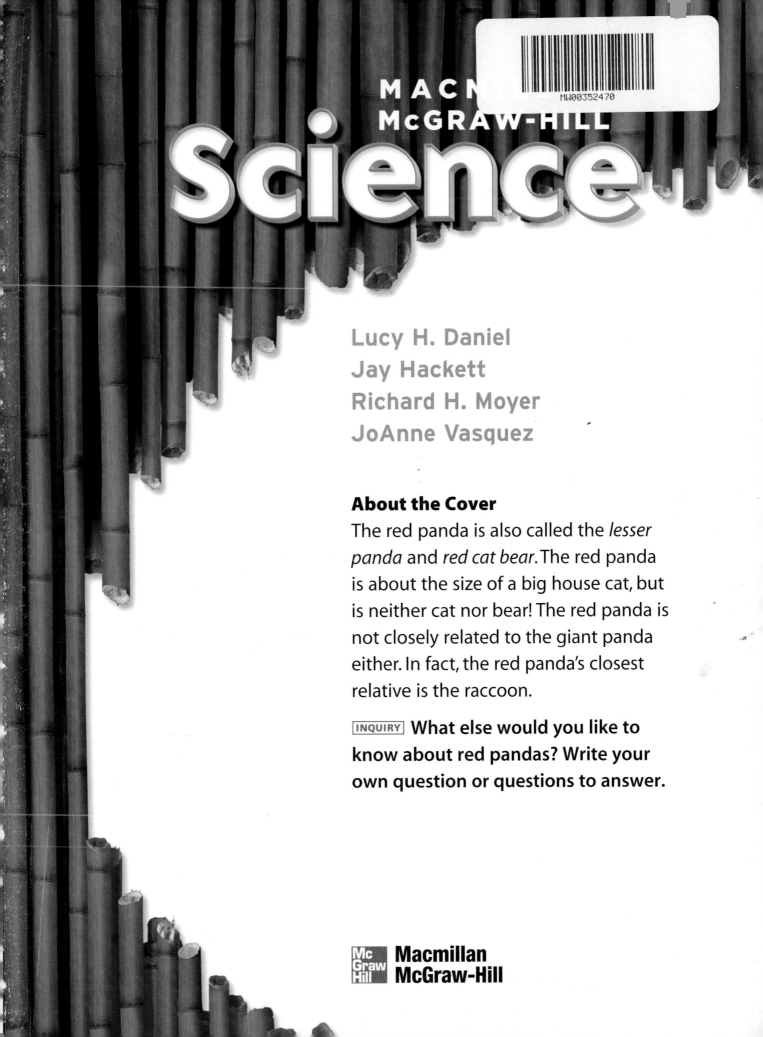

MACN
McGRAW-HILL
Science

Lucy H. Daniel

Jay Hackett

Richard H. Moyer

JoAnne Vasquez

About the Cover

The red panda is also called the *lesser panda* and *red cat bear*. The red panda is about the size of a big house cat, but is neither cat nor bear! The red panda is not closely related to the giant panda either. In fact, the red panda's closest relative is the raccoon.

INQUIRY What else would you like to know about red pandas? Write your own question or questions to answer.

Mc Graw Hill **Macmillan McGraw-Hill**

Program Authors

Dr. Lucy H. Daniel
Teacher, Consultant
Rutherford County Schools, North Carolina

Dr. Jay Hackett
Professor Emeritus of Earth Sciences
University of Northern Colorado

Dr. Richard H. Moyer
Professor of Science Education
University of Michigan-Dearborn

Dr. JoAnne Vasquez
Elementary Science Education Consultant
Mesa Public Schools, Arizona
NSTA Past President

Contributing Authors

Lucille Villegas Barrera, M.Ed.
Elementary Science Supervisor
Houston Independent School District
Houston, Texas

Mulugheta Teferi, M.A.
St. Louis Public Schools
St. Louis, Missouri

Dinah Zike, M.Ed.
Dinah Might Adventures LP
San Antonio, Texas

The features in this textbook entitled "Amazing Stories," as well as the unit openers, were developed in collaboration with the National Geographic Society's School Publishing Division.

Copyright © 2002 National Geographic Society. All rights reserved.

learning through listening

Students with print disabilities may be eligible to obtain an accessible, audio version of the pupil edition of this textbook. Please call Recording for the Blind & Dyslexic at 1-800-221-4792 for complete information.

The McGraw·Hill Companies

Macmillan McGraw-Hill

Published by Macmillan/McGraw-Hill, of McGraw-Hill Education, a division of The McGraw-Hill Companies, Inc., Two Penn Plaza, New York, New York 10121.

Printed in the United States of America

ISBN 0-02-282593-2

5 6 7 8 9 110/043 09 08 07

Teacher Reviewers

Michelle Dunning
Birmingham, Alabama

Donna Bullock
Chandler, Arizona

Debra Allen
Davie, Florida

Lora Meade
Plantation, Florida

Roxanne Laird
Miami, Florida

Karen Gaudy
Satellite Beach, Florida

Stephanie Sirianni
Margate, Florida

Heidi Stephens
South Daytona, Florida

Rosanne Phillips
Miami, Florida

Brenda Crow
Miami, Florida

Kari Pingel
Pella, Iowa

Christie Jones
Springfield, Illinois

Diane Songer
Wabash, Indiana

Lee Arwood
Wabash, Indiana

Margarite Hart
Indianapolis, Indiana

Charlotte Bennett
Newburgh, Indiana

Donna Halverson
Evansville, Indiana

Stephanie Tanke
Crown Point, Indiana

Mindey LeMoine
Marquette, Michigan

Billie Bell
Grand View, Missouri

Charlotte Sharp
Greenville, North Carolina

Pat Shane
Chapel Hill, North Carolina

Karen Daniel
Chapel Hill, North Carolina

Linda Dow
Concord, North Carolina

Life Science

Consultants

Dr. Carol Baskin
University of Kentucky
Lexington, KY

Dr. Joe W. Crim
University of Georgia
Athens, GA

Dr. Marie DiBerardino
Allegheny University of
Health Sciences
Philadelphia, PA

Dr. R. E. Duhrkopf
Baylor University
Waco, TX

Dr. Dennis L. Nelson
Montana State University
Bozeman, MT

Dr. Fred Sack
Ohio State University
Columbus, OH

Dr. Martin VanDyke
Denver, CO

Dr. E. Peter Volpe
Mercer University
Macon, GA

Earth Science

Consultants

Dr. Clarke Alexander
Skidaway Institute of
Oceanography
Savannah, GA

Dr. Suellen Cabe
Pembroke State University
Pembroke, NC

Dr. Thomas A. Davies
Texas A & M University
College Station, TX

Dr. Ed Geary
Geological Society of America
Boulder, CO

Dr. David C. Kopaska-Merkel
Geological Survey of Alabama
Tuscaloosa, AL

Physical Science

Consultants

Dr. Bonnie Buratti
Jet Propulsion Lab
Pasadena, CA

Dr. Shawn Carlson
Society of Amateur Scientists
San Diego, CA

Dr. Karen Kwitter
Williams College
Williamstown, MA

Dr. Steven Souza
Williamstown, MA

Dr. Joseph P. Straley
University of Kentucky
Lexington, KY

Dr. Thomas Troland
University of Kentucky
Lexington, KY

Dr. Josephine Davis Wallace
University of North Carolina
Charlotte, NC

Consultant for Primary Grades

Donna Harrell Lubcker
East Texas Baptist University
Marshall, TX

Teacher Reviewers (continued)

Beth Lewis
Wilmington, North Carolina

Cindy Hatchell
Wilmington, North Carolina

Cindy Kahler
Carrborro, North Carolina

Diane Leusky
Chapel Hill, North Carolina

Heather Sutton
Wilmington, North Carolina

Crystal Stephens
Valdese, North Carolina

Meg Millard
Chapel Hill, North Carolina

Patricia Underwood
Randleman, North Carolina

E. Joy Mermin
Chapel Hill, North Carolina

Yolanda Evans
Wilmington, North Carolina

Tim Gilbride
Pennsauken, New Jersey

Helene Reifowitz
Nesconsit, New York

Tina Craig
Tulsa, Oklahoma

Deborah Harwell
Lawton, Oklahoma

Kathleen Conn
West Chester, Pennsylvania

Heath Renninger Zerbe
Tremont, Pennsylvania

Patricia Armillei
Holland, Pennsylvania

Sue Workman
Cedar City, Utah

Peg Jensen
Hartford, Wisconsin

Life Science

UNIT B

Where Plants and Animals Live PAGE B1

Science Handbook

Health Handbook

Activities

Unit B

Explore Activities

Quick Labs with FOLDABLES™

Inquiry Skill Builders

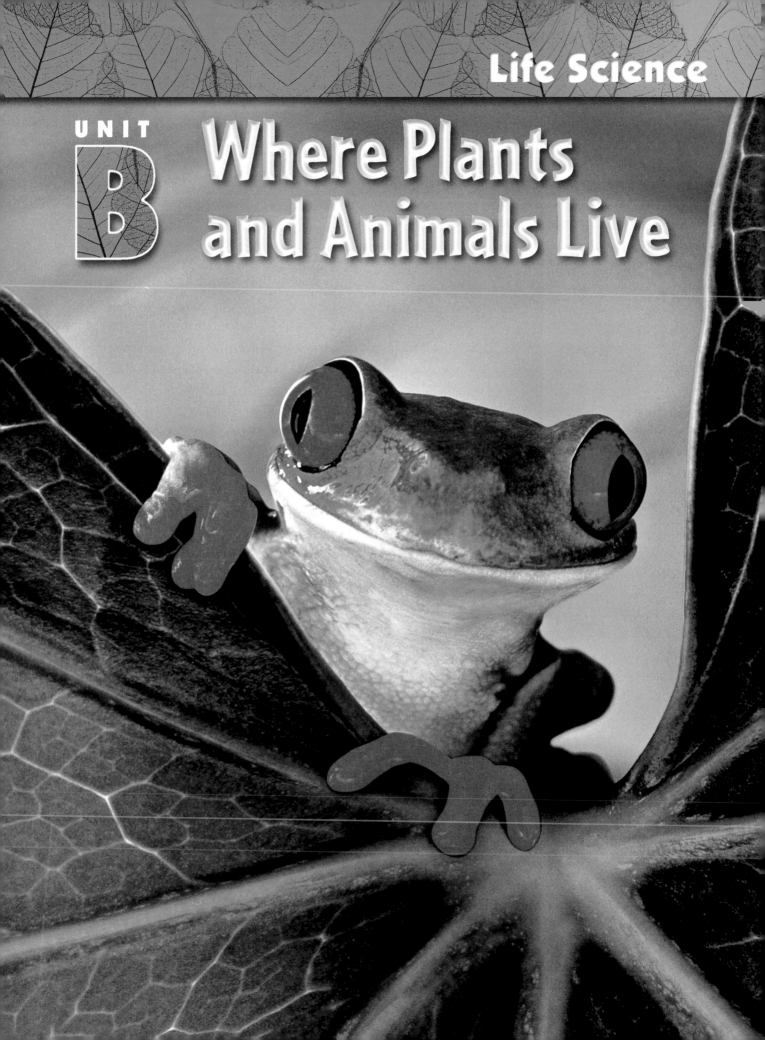

Life Science

UNIT B
Where Plants and Animals Live

LOOK!

The red-eyed tree frog can climb trees. What body parts help it climb?

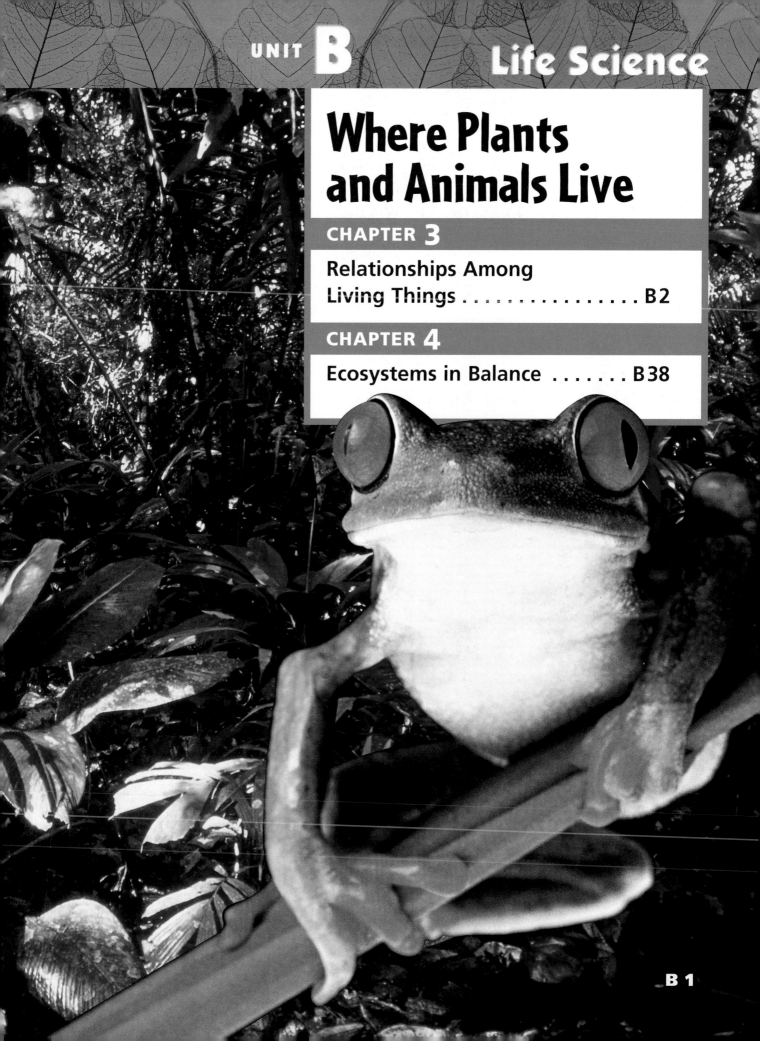

Where Plants and Animals Live

Relationships Among Living Things

Did You Ever Wonder?

What kinds of plants and animals live in the Chesapeake
Bay? Recently people have been helping to protect this
environment. As a result the numbers of plants and animals
that live here are increasing.

INQUIRY SKILL Communicate What can you do to help the
living things in your environment?

Ecosystems

Get Ready

Do you think a caribou could survive in a desert? Caribou live in the cold Arctic tundra. Their thick fur keeps them warm. Their strong hooves help them walk over the ground. Look at the picture of the caribou. What are their surroundings like? Why do you think they live where they do?

Inquiry Skill

You **observe** when you use one or more of the senses to learn about an object or event.

B 4

Explore Activity

What Can You Find in an Ecosystem?

Materials

meter tape

ball of yarn

4 clothespins

hand lens

Procedure

1 **Measure** Mark off an area of ground that is 1 meter square. Stick a clothespin into the ground at each corner. Wrap yarn around the tops of the clothespins.

2 **Observe** Use your hand lens to look at the living and nonliving things in this area.

3 Use a chart to record what you see. Label each object *living* or *nonliving*.

4 Share your findings with a classmate. Compare the environments each of you observed.

Drawing Conclusions

1 How many different kinds of nonliving things are in your environment? What did you have the most of?

2 Choose one living thing you observed. What are the characteristics of this organism?

3 FURTHER INQUIRY Infer What are the characteristics of another living thing that might live here? How do you know?

Read to Learn

Main Idea Ecosystems are made up of living and nonliving things.

A population of zebras live on the grassland.

What Makes Up an Ecosystem?

Plants and animals live with one another. They also depend on one another to stay alive. They depend on nonliving things, too, like rocks, soil, water, and air. Together, all the living and nonliving things in a place make up an **ecosystem** (EK·oh·sis·tuhm). Ecosystems can be large or small. Earth can be thought of as one large ecosystem. An ecosystem can also be as small as the space under a rock.

All the living things in an ecosystem make up a **community** (kuh·MYEW·ni·tee). The community on the grasslands includes zebras, giraffes, and other animals, along with trees and grasses.

Each community is made up of many different **populations** (pahp·yuh·LAY·shuhnz). A population is all the members of a single type of organism. All the giraffes on a grassland make up the giraffe population. All the zebras make up the zebra population.

Different populations make up a community.

B 6

A tree is one part of a forest ecosystem. It is also home to birds, insects, and squirrels. For these animals the tree is a **habitat** (HAB·i·tat). A habitat is a living thing's home. Living things get food, water, and shelter from their habitat.

The different parts of an ecosystem affect one another. Insects live in the tree's bark. Birds build nests from twigs. The tree grows in soil and takes in water and air.

Some parts of an ecosystem are so small you cannot see them! Tiny organisms live in the soil. They break down materials such as leaves and add them to the soil.

READING Summarize
What are some of the living and nonliving things in an ecosystem?

1 Leaves fall from the tree during autumn.

2 Small organisms in the soil break down the leaves.

3 Leaves become part of the soil.

4 The tree grows in the soil.

What Habitats Are Found in a Pond?

A frog needs food, water, and shelter. It gets all of these things at a pond. The pond is the frog's habitat.

Some plants and animals live only in parts of the pond. A sunfish stays in the water. Raccoons live along the banks. Lily pads float on the water's surface. The pond provides many different habitats.

A pond is an ecosystem. A pond is also a *wetland*. Wetlands are a mix of land and water ecosystems. Marshes and swamps are also wetlands. Wetlands are important because they provide homes for many different types of plants and animals.

▷ **How can a pond provide different habitats?**

READING
Diagrams

1. What are two habitats shown in the diagram?
2. Choose three of the living things that live in a pond. Make a chart that shows each organism's habitat.

Habitats

❶ The Banks
Plants like ferns and mosses live along the pond's banks. Animals include insects, mice, snakes, raccoons, and birds.

❷ The Water's Edge
Plants on the water's edge live partly underwater. Animals include salamanders, snails, and water bugs.

❸ Shallow Water
Floating plants live in the shallow water. Frogs sit on lily pads, and turtles lie on rocks to bask in the sunlight.

❹ Deep Water
Floating plants live here, too. Fish live in the deep water.

A Pond Ecosystem

Populations

5 Great Blue Heron
The heron flies from place to place. It comes to the pond for its favorite foods.

6 Dragonfly
Young dragonflies live in the water. They breathe through gills. Adult dragonflies live out of the water.

7 Bladderwort
The bladderwort is a floating plant. Its stems and leaves have air sacs. When a small organism touches an air sac, it is sucked inside and eaten!

8 Algae
Algae are tiny organisms. A pond may contain billions of algae.

Inquiry Skill
BUILDER

What Makes Up a Forest Community?

You have learned that a community is all the living things in an ecosystem. Different ecosystems have different communities. For example, the pond community includes bladderworts, frogs, algae, and dragonflies. What makes up a forest community? Look at the picture on this page. Use your observations to define a forest community.

Procedure

1. **Observe** Make a list of all the things you see in the picture of the forest.

2. **Classify** Which of the things on your list are living? Which are nonliving?

Drawing Conclusions

Define Terms Using your list, explain what a forest community is.

Why It Matters

People can help habitats. Many people are carefully watching habitats. They look for changes. When they find a problem, they try to solve it. For example, a plant disease may have killed many plants in an area. Scientists may find ways to help plants produce seeds and grow back.

e-Journal Visit our Web site www.science.mmhschool.com to do a research project on ecosystems.

Think and Write

1. What is an ecosystem?

2. What are some members of a pond community?

3. How is a community different from a population?

4. **INQUIRY SKILL** **Define Terms** Think about the habitat of a pet dog. List some of the characteristics of the habitat. Use your list to write your own definition of *habitat*.

5. **Critical Thinking** How would a pond ecosystem change if the water in the pond dried up?

L·I·N·K·S

MATH LINK

Solve a problem. An ecosystem has frogs, dragonflies, great blue herons, and floating plants. How many animals are in the community? How many populations are there? Use the graph below.

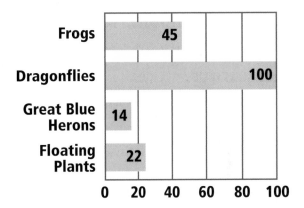

Frogs	45
Dragonflies	100
Great Blue Herons	14
Floating Plants	22

0 20 40 60 80 100

WRITING LINK

Explanatory Writing Make a list of different insects that live in your neighborhood. Tell how insects help or use other living things.

SOCIAL STUDIES LINK

Make a map. Make an outline map of your state. Draw the major bodies of water. Include mountains, rivers, and any other important features. Label your map.

TECHNOLOGY LINK

LOG ON Visit www.science.mmhschool.com for more links.

You Depend on Plants and Animals!

What would your life be like without plants and animals? It would be different. You depend on plants and animals in many ways.

Do you ever wear cotton clothes? Cotton is a plant fiber. Are you sitting on a wooden chair or holding a wooden pencil? Wood comes from trees. The trunks and branches of trees are wood. Do you like eating fruits and vegetables? If so, you like to eat plants.

You even use plants while you breathe! Every time you breathe in, you take in a gas called oxygen that comes from plants.

How do you depend on animals? Animals provide meat and other foods. Milk, ice cream, cheese, and eggs all come from animals.

How Cloth Is Made

1. Cotton is stripped or picked from the plants.

2. Cotton fibers are separated from cotton seeds.

3. Dirt is removed from the fibers.

4. The fibers are spun into yarn.

5. The yarn is woven into cloth.

6. The cloth is cut and sewn to make clothing.

Do you own a wool sweater or mittens? Wool is usually made from a sheep's soft, curly fleece.

Leather is made from animal skins. Shoes, gloves, and wallets are often made of leather.

Some people depend on animals to help them get around. Some animals can be trained to be great helpers—and pals.

Cows provide people with milk.

What Did I Learn?

1. Which of the following is not provided by trees?

   A wooden chair

   B pencil

   C wool

   D paper

2. What step comes after cotton fibers are separated from cotton seeds?

   F The fibers are spun into yarn.

   G The yarn is woven into cloth.

   H The cloth is cut and sewn into clothing.

   J Dirt is removed from the fibers.

LOG ON Visit www.science.mmhschool.com to learn more about plants and animals.

Food Chains and Food Webs

Vocabulary

producer, B16

consumer, B17

food chain, B17

decomposer, B18

food web, B20

energy pyramid, B22

Get Ready

Would you eat grass for lunch? No, you would not. For a donkey, however, grass makes a fine meal. Cows, horses, sheep, and buffalo eat grass, too! Other animals, like eagles and lions, never eat grass. Different animals eat different things. What kinds of foods do you eat?

Inquiry Skill

You **classify** when you group similar things together.

B 14

Explore Activity

Where Does Food Come From?

Procedure

1 **Observe** Look at the picture of the pizza. What types of foods do you see? Make a list.

2 **Classify** Next to each thing on your list, record whether the food comes from a plant or an animal. Write *P* for plant and *A* for animal.

3 Look at your list of foods that come from animals. From which animal does each food come? What food does that animal eat?

Drawing Conclusions

1 If there were no plants, which foods would be left to make pizza? (Hint: Think about what animals eat to survive.)

2 **Infer** Do all foods come from plants? Explain your answer.

3 Write down a food that you like to eat. List all of the ingredients in this food.

4 FURTHER INQUIRY **Classify** Make a chart that shows where each ingredient in your favorite food comes from.

Main Idea Animals depend on plants for their food.

The plant is a producer.

What Makes Up a Food Chain?

What is food? Food is material that organisms use to get energy. All food comes from organisms called **producers** (pruh·DEW·suhrz). Producers make food from water, air, and energy from sunlight. Green plants and some one-celled organisms are producers.

The insect is a consumer.

Desert Food Chain

Plant makes food   **Insect eats plant**   **Mouse eats insect**

Ocean Food Chain

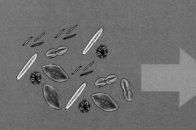

Algae make food   **Tiny consumers eat algae**   **Shellfish eats tiny consumers**

Animals are **consumers** (kuhn·SEW·muhrz). Consumers are organisms that eat producers or other consumers.

Together, producers and consumers make up a **food chain**. A food chain is a series of organisms that depend on one another for food. Food chains start with producers. Consumers eat those producers. Other consumers eat the first consumers.

▷ **What is an example of a producer? A consumer?**

READING
Diagrams

1. What are the first and last organisms in the desert food chain? In the ocean food chain?

2. How are the two food chains different? How are they alike?

Snake eats mouse

Hawk eats snake

Small fish eat shellfish

Large fish eats small fish

Killer whale eats large fish

How Are Materials Recycled?

What do you think happens to a leaf that falls from a tree? Over time organisms break the leaf apart and return it to the soil. These kinds of organisms are called **decomposers** (dee·kuhm·POH·zuhrz). A decomposer breaks down dead plant and animal material. It recycles chemicals so that they can be used again.

Producers, consumers, and decomposers work together to recycle materials through an ecosystem. Producers use the recycled material to make new food. Consumers eat the food. When producers and consumers die, decomposers recycle the dead material. The cycle goes on and on.

This cycle helps reduce garbage. Without decomposers Earth would be covered with dead plant and animal material. When decomposers eat plant and animal material, they make compost. Compost is a mix of decaying leaves, vegetables, and other living matter.

Decomposers are breaking down leaves. The leaves will become part of the soil.

Decomposers include fungi. You can see fungi growing on this tree.

On a cold morning, you might see a compost pile steaming. That is because this process gives off heat. Compost piles heat up for the same reason that you heat up when you exercise. Like your muscles, the decomposers are working hard and are using lots of fuel. When they use lots of fuel, heat is produced.

READING **Summarize**

What is the job of a decomposer?

How does composting help recycle plant and animal material?

QUICK LAB

Decomposers

FOLDABLES Make a Folded Table. (See p. R 44.) Lable the table as shown.

BE CAREFUL! Don't open the sealed bag.

Day 1	
Day 2	
Day 3	
Day 4	
Day 5	
Day 6	
Day 7	

1. Put some apple pieces in a plastic bag. Seal the bag.

2. **Observe** Leave the apples in the bag for one week. Observe the apples every day. Record your observations in the table.

3. What happened to the pieces of apple? Record your findings on the back of the table.

4. What does this activity tell you about decomposers? Write your answer on the back of the table.

5. **Infer** What would happen if there were no decomposers? Write your answer on the back of the table.

What Is a Food Web?

An owl and a hawk are not in the same food chain. They are still connected, however. Both eat many of the same things, like mice and snakes.

In the desert owls and hawks are parts of the same **food web**. A food web is made up of several food chains that are connected.

Look at the single desert food chain on page B16. Now compare it with the food web shown below. Notice how the food chains connect to one another. Try to find the chain with grass, jackrabbit, and coyote.

Herbivore
An animal that eats only plants is called a *herbivore*. Jackrabbits and prairie dogs are herbivores.

Carnivore
An animal that eats only other animals is called a *carnivore*. Hawks, snakes, owls, and coyotes are carnivores.

Different consumers eat different kinds of food. Some consumers eat only plants. Some eat only animals. Still others eat both plants and animals. Each group has a special name. This name ends with the letters *vore*. *Vore* means "eater."

▶ **What groups of animals are part of a food web?**

READING Diagrams

1. What are three different food chains in the food web?

2. If the jackrabbits left the desert, how would the food web change?

Omnivore
An animal that eats both plants and animals is called an *omnivore*. Javelinas, small piglike animals, are omnivores. They eat insects, cacti, and other desert plants.

What Is an Energy Pyramid?

You can group organisms by their positions in food webs. Each group forms its own level on an **energy pyramid** (EN·uhr·jee PIR·uh·mid). An energy pyramid is a diagram that shows how energy moves through an ecosystem.

Each level in an energy pyramid has more members than the level above it. There are more producers than plant eaters. There are more plant eaters than meat eaters. For every bald eagle, there might be hundreds of insects and thousands of plants.

▷ **Are there more producers or consumers in an ecosystem?**

READING
Diagrams

How do the levels compare in an energy pyramid?

Energy Pyramid

Animals that are not hunted by other animals

Animals that eat other animals

Animals that eat plants

Plants

Why It Matters

You are a consumer. You are part of a food web. You probably depend on many different producers as well as other consumers for food. You also depend on decomposers to recycle plant and animal materials in your ecosystem.

e-Journal Visit our Web site www.science.mmhschool.com to do a research project on food chains.

Think and Write

1. Where does food come from?

2. What is a producer? A consumer? A decomposer?

3. How are food webs different from food chains?

4. How can composting be used to recycle discarded plant and animal material? Explain your answer.

5. **Critical Thinking** What would happen if an ecosystem had more consumers than producers? Could this ecosystem last? Why?

L·I·N·K·S

WRITING LINK

Writing a Story Create a picture book. Write a story about the food web. Include yourself as a character. Remember to have a beginning, a middle, and an end to the story. Include a diagram of the food web in your story.

MATH LINK

Solve a problem. Eli is collecting fresh vegetables for a pizza. He picks 14 tomatoes, 6 peppers, and 3 onions. How many more tomatoes than peppers did Eli pick? Explain how you found your answer.

SOCIAL STUDIES LINK

Use a map. Petrified Forest National Park is located in Arizona. Use a map to find out where in the state it is located. Research and list the plants and animals that live there. Find out how this ecosystem has been affected by humans.

TECHNOLOGY LINKS

Science Newsroom CD-ROM Choose *Chains of Life* and *Down to Earth* to learn more about food chains and food webs.

 LOG ON Visit www.science.mmhschool.com for more links.

Roles for Plants and Animals

Vocabulary

carbon dioxide and oxygen cycles, B27

predator, B28

prey, B28

scavenger, B29

parasite, B31

host, B31

Get Ready

What do these fish need to survive? What do these plants need to survive? Do you think they might need each other? What would you do to take care of the living things in an aquarium?

Inquiry Skill

You make a model when you make something that represents objects or events.

Explore Activity

How Do Living Things Meet Their Needs?

Materials
gravel

guppy or goldfish

small water plants

2-L plastic drink bottle

bottom of another drink bottle with holes

fish food

Procedure

BE CAREFUL! Handle animals carefully. Measure materials carefully.

1 Make a Model Put a 3-cm layer of gravel into the plastic drink bottle. Fill the bottle with water as shown. Anchor the plants in the gravel.

2 Cover the bottle with the bottom of another bottle. Do not place it in direct sunlight.

3 After two days, gently place the fish in the bottle. Add a few flakes of fish food.

4 Observe Look at your ecosystem every day for two weeks. Feed the fish twice each week. Record your observations.

Drawing Conclusions

1 What did the fish need to survive? What did the plants need to survive?

2 What might happen if the plant was not part of the ecosystem?

3 **FURTHER INQUIRY** **Experiment** How do frogs meet their needs? Predict what kind of ecosystem you would need to build to find out. How do you know?

Main Idea Living things depend on one another in many ways.

How Do Living Things Use Air?

What do you need from the air around you? You can't see it, but you take it in with every breath. It's a gas called oxygen. All animals need oxygen. Animals that live in water get their oxygen from the water.

Plants need gases from the air, too. They need carbon dioxide to make food. They also need oxygen to use food. During the day plants make their own food. At night they get oxygen from the air, just like animals.

Where do these gases come from? They come from plants and animals! Plants make oxygen, a gas that animals need. Animals give off carbon dioxide, a gas that plants need.

Trees give off oxygen.

Trees take in carbon dioxide.

Plants give off oxygen.

Fish take in oxygen.

Plants take in carbon dioxide.

Fish give off carbon dioxide.

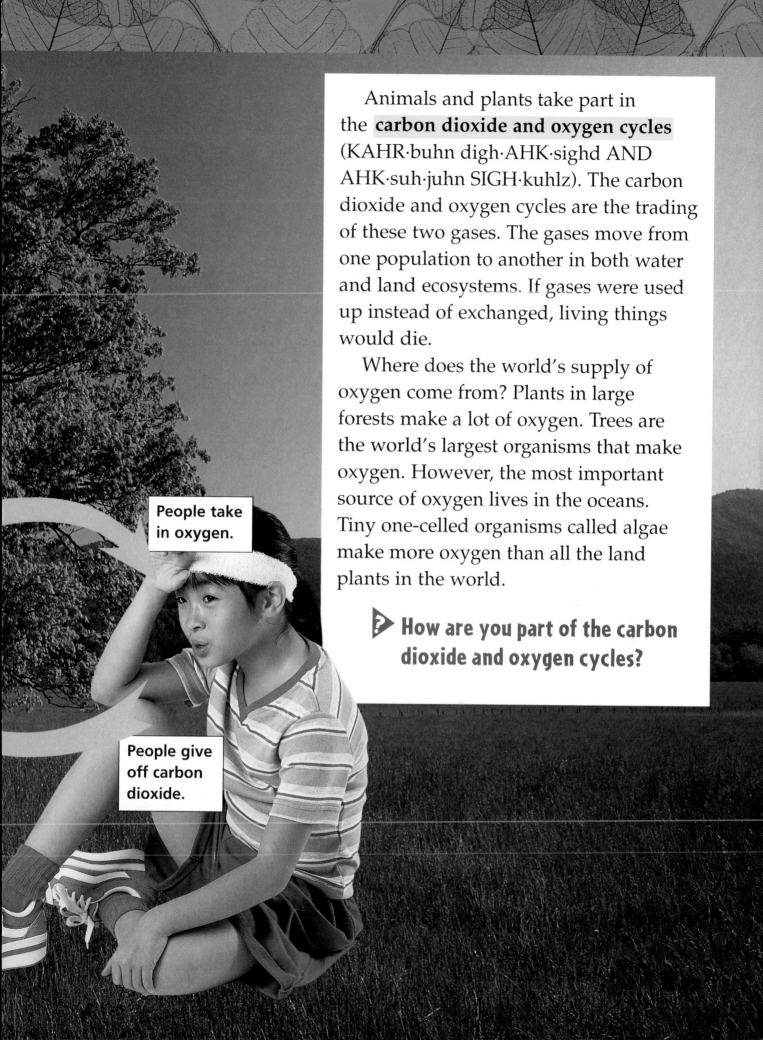

Animals and plants take part in the **carbon dioxide and oxygen cycles** (KAHR·buhn digh·AHK·sighd AND AHK·suh·juhn SIGH·kuhlz). The carbon dioxide and oxygen cycles are the trading of these two gases. The gases move from one population to another in both water and land ecosystems. If gases were used up instead of exchanged, living things would die.

Where does the world's supply of oxygen come from? Plants in large forests make a lot of oxygen. Trees are the world's largest organisms that make oxygen. However, the most important source of oxygen lives in the oceans. Tiny one-celled organisms called algae make more oxygen than all the land plants in the world.

▷ **How are you part of the carbon dioxide and oxygen cycles?**

People take in oxygen.

People give off carbon dioxide.

This gecko will have a tasty treat.

What Do Populations Depend on?

Visit a forest, and you may see animals eating. Mice and rabbits eat grasses and shrubs. Beetles eat leaves. Owls and hawks swoop down from the sky to catch mice. Populations depend on the resources in the community.

Owls, coyotes, wolves, and many other animals are hunters. Animals that hunt for food are called **predators** (PRED·uh·tuhrz). The animals that predators eat are called **prey** (PRAY).

Predators have body parts that allow them to hunt and catch prey. Some are able to run very fast. Many have a keen sense of smell or sight.

A heron catches a fish.

A lion cub begins to hunt at the age of six months.

Do prey depend on predators? You might not think so. However, predators help control the populations of many animals. If predators disappeared, the numbers of their prey would rise quickly. After the prey ate all the food that was available, many of them would die.

Not all meat eaters kill what they eat. Have you ever seen a crow pick at a dead animal? Crows are **scavengers** (SKAV·uhn·juhrz). Scavengers eat dead animals. Crayfish, crabs, vultures, and many other animals are scavengers.

▷ **Why do prey need predators?**

Vultures

A hyena is a scavenger. Why are scavengers important?

This snake is eating an egg.

The oxpecker and the buffalo help each other.

Cows help cattle egrets find food.

How Can Populations Affect Each Other?

To many fish, sea anemones are very dangerous. Their tentacles contain a strong poison. However, the clownfish can swim near an anemone without being harmed! Its body is coated with special slime that protects it.

The clownfish and anemone help each other. When a clownfish feels threatened, it swims to the anemone's tentacles for safety. The anemone feeds on scraps that fall out of the clownfish's mouth.

How do the sea anemone and clownfish help each other?

The buffalo and oxpecker also help each other. An oxpecker pecks insects out of the skin of a buffalo. It also calls out loudly when it sees danger. This warns the buffalo.

Sometimes one population helps another without being helped in return. Cows help cattle egrets in this way. Egrets follow cows wherever the cows go. The cows are so big that they stir up insects and other small animals. The egrets eat these insects and animals. The cows help the egrets find food but receive no help in return. They are not harmed, either.

Sometimes one population does harm another. Look at the tapeworm. It is a **parasite** (PAR·uh·sight). A parasite is an organism that lives on or inside another organism. The organism a parasite lives with is called the **host** (HOHST).

A tapeworm attaches itself inside the host. The host can be a human or another animal. The tapeworm takes in food that the host has digested. Parasites like tapeworms can make their host sick. Sometimes they kill their host.

Fleas are parasites on dogs. They live on a dog's skin and take in its blood. Fleas harm dogs but rarely kill them.

A tapeworm is a parasite.

▷ **What is a parasite?**

Fleas are parasites. They live on cats and dogs.

Traveling Seeds

Make a Folded Chart. (See p. R44.) Label the chart as shown.

Prediction	Results

1. **Predict** What will happen when you toss seeds onto a piece of fake fur? Record your prediction on the chart.

2. **Experiment** Test your prediction. Have your partner hold up the fur. Toss different seeds at it. Record the results on the chart.

3. Which seeds stuck to the fur? Were your predictions correct? Write your answer on the back of your Foldables chart.

How Do Animals Help Plants Reproduce?

An oak tree makes seeds called acorns. Acorns might grow into new oak trees if they land in the right place. Most acorns fall right under the tree. Is this a good place for acorns to grow? No, it is not. There is not enough light.

How can acorns be moved? Animals help! Squirrels find the acorns. They bury them in the ground to store them for winter. They eat many of the acorns, but a few acorns are forgotten. They stay buried in the ground, far from the tree. They might grow into new trees.

READING Summarize
How do squirrels help oak trees reproduce?

Lesson Review

Why It Matters

Some organisms can be both helpful and harmful. Bacteria are just such an organism. They are one-celled living things. They are helpful because they are decomposers. Decomposers break down dead plant and animal matter. Decomposers help recycle Earth's materials.

e-Journal Visit our Web site www.science.mmhschool.com to do a research project on prey and predators.

Think and Write

1. What are the carbon dioxide and oxygen cycles?

2. What are predators and prey?

3. Give an example of two organisms that help each other survive.

4. How do squirrels help oak trees reproduce?

5. Critical Thinking Most parasites harm their host but do not kill it. Why does a parasite need a living host?

L·I·N·K·S

MATH LINK

Make a graph. A shark can smell its prey 400 meters (440 yards) away. It can hear sounds about 910 meters (1,000 yards) away. Make a bar graph to show this information.

WRITING LINK

Expository Writing Write a list of questions you would ask if you could interview a predator. Research the animal to find the answers. With a partner, record the interview. Play the recording for the class.

LITERATURE LINK

Read *Kit Foxes* to learn how two kit foxes find a den and raise a family. When you finish reading, think about an animal that might live in your neighborhood. Try the activities at the end of the book.

Kit Foxes
Written by Jennifer Jacobson • Illustrated by Pat Traub

TECHNOLOGY LINK

LOG ON Visit www.science.mmhschool.com for more links.

LONG LIVE THE RAIN FORESTS

Where can you find a monkey small enough to fit it your hand? Where can you find a frog that can soar like a bird? Or find a flower the size of a chair? These treasures live in the world's tropical rain forests. Rain forests are wet, warm lands that are home to many living things.

Millions of kinds of plants and animals live in rain forests. The plants help us because they make oxygen, an important gas in the air. Rain forest plants also provide many of the world's medicines.

Sadly, large chunks of rain forest are destroyed every minute. The forests are being destroyed for many reasons. Loggers cut down trees and sell the lumber. Farmers clear land for cattle and crops. Miners dig up the land to take the minerals below. More than half of the world's rain forests are gone already.

Banana tree

This Blue Morpho butterfly is one of millions of insects that live in the rain forest.

People around the world now understand why rain forests are important. People are working hard to save them. You can help, too. Learn more about rain forests. Tell your friends and the adults you know what you find out.

Write ABOUT IT

1. Why do people destroy rain forests?
2. Do you think rain forests are worth saving? Why or why not?

LOG ON Visit www.science.mmhschool.com to learn more about rain forests.

Chapter 3 Review

Vocabulary

Fill each blank with the best word from the list.

community, B6 **parasite,** B31

consumer, B17 **predator,** B28

decomposer, B18 **prey,** B28

habitat, B7 **producer,** B16

host, B31 **scavenger,** B29

1. The place where an animal lives and grows is its _____.

2. All the living things in an ecosystem make up a(n) _____.

3. Bacteria are a kind of _____ because they break down dead plant and animal material.

4. A crow is a(n) _____ because it eats dead animals.

5. A _____	hunts for	**6.** _____.
7. A _____	lives on or in a(n)	**8.** _____.
9. A _____	eats foods made by a(n)	**10.** _____.

Test Prep

11. Plants and animals exchange carbon dioxide and _____.

A food

B oxygen

C water

D molds

12. All the following are consumers in a pond EXCEPT _____.

F algae

G herons

H frogs

J dragonflies

13. Animals can help plants by _____.

A making food

B finding water

C spreading seeds

D providing energy

14. Bacteria are helpful because they _____.

F produce food

G spoil food

H are parasites

J recycle materials

15. A parasite depends on a

_____.

A scavenger

B producer

C host

D decomposer

Concepts and Skills

16. Reading in Science Use this diagram to explain a food chain. Write a short paragraph that explains your answer.

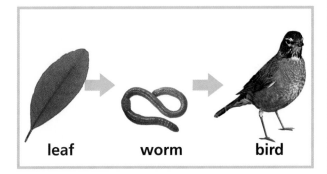

leaf   worm   bird

17. Product Ads Some insects are pests. Some ads claim their products are strong enough to kill many insects. Write down ways these products might affect a food chain.

18. Decision Making The sign says, "Do not feed the ducks." You brought some bread to the pond. Would you feed the ducks? Write a paragraph explaining your decision.

19. Critical Thinking A pack of coyotes moves to a new ecosystem. They eat many rabbits. What can happen to the ecosystem? Explain your answer in writing.

20. INQUIRY SKILL **Define Terms Based on Observations** Using your observations, write a definition of an aquarium ecosystem. Describe the parts of the ecosystem and how they work together.

Did You Ever Wonder?

INQUIRY SKILL **Communicate** A habitat is an organism's home. It provides food and shelter, as well as protection. How might people help to conserve the places where organisms live?

LOG ON Visit **www.science.mmhschool.com** to boost your test scores.

Ecosystems in Balance

Did You Ever Wonder?

What fish lives in both oceans and rivers? The answer is the salmon. Salmon are born in rivers, then move to oceans. At the end of their lives, they swim up the same river in which they were born. The journey is dangerous!

INQUIRY SKILL Infer Why do you think salmon return to the river?

Competition Among Living Things

Get Ready

Who gets the bug? All of the young birds want it. Plants and animals compete for their basic needs. One of these needs is space. Does the amount of space available affect the way plants grow?

Inquiry Skill

You infer when you form an idea from facts or observations.

Explore Activity

How Much Room Do Plants Need?

Materials

soil

bean seeds

4 milk cartons

measuring cup

water

masking tape

marker

Procedure

1 Cut the tops from the milk cartons. Use the masking tape and the marker to label the cartons A to D. Use the measuring cup to fill each carton with the same amount of soil.

2 **Use Variables** Plant 3 bean seeds in carton A. Plant 6 bean seeds in carton B. Plant 12 bean seeds in carton C and 24 bean seeds in carton D.

3 **Predict** What do you think each carton will look like in 14 days? Record your predictions.

4 **Experiment** Place the cartons in a well-lighted area. Water the plants every two days. Use the same amount of water for each carton. Record any changes you observe in the plants.

Drawing Conclusions

1 How do the plants in carton D compare with the plants in the other cartons?

2 What are the plants competing for?

3 Repeat this activity. Compare your results. What happened?

4 FURTHER INQUIRY **Infer** How do plants in your neighborhood compete for what they need to grow?

Main Idea Plants and animals compete for the things they need.

How Much Room Do Organisms Need?

The owl is a predator. The mouse is its prey.

Competition (kahm·pi·TISH·uhn) for space affects how plants grow. Competition occurs when organisms with similar needs work against one another to get what they need to live. Organisms share and compete for space, water, food, and oxygen.

Desert plants compete for water. A cactus soaks up all of the moisture in a single area. No other plants can grow in this area.

Rabbits compete for food. If there are too many rabbits, all of the grass will be eaten. Some of the rabbits won't survive.

Predators also compete. Predators are animals that hunt for food. Predators compete for prey. *Prey* try to escape from predators. Predators compete with each other. Hawks, owls, and snakes all compete to catch mice for food.

These cacti are competing for water. How does that explain why they grow so far apart?

Different ecosystems support different numbers of organisms. The chart below compares the numbers of different types of trees and birds in a rain forest with those in a temperate (TEM·puhr·it) forest. Temperate forests are the most common type of forest in the United States. Rain forests support many more types of trees and birds than temperate forests.

The same pattern is true for other living things. Why are there so many more types of organisms in the rain forest? Rain forests are much wetter and warmer than temperate forests. More plants can grow in this environment. More plants means more animals, too.

Temperate forest

READING **Compare and Contrast**
How are rain forests and temperate forests alike? How are they different?

Comparing Forests		
	Different Types of Trees	**Different Types of Birds**
Temperate forest	50 to 60	50
Rain forest	500 to 600	250

Rain forest

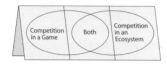

Can Competition Be Avoided?

Competition is a struggle for survival. In order to survive, some organisms find ways to avoid competing. Many types of organisms share the same ecosystem. Each type of organism has its own **niche** (NICH). A niche is the job or role an organism has in an ecosystem. An organism's niche includes what an organism does, what it eats, and how it interacts with other organisms.

For example, there are many types of pigeons in the forests of New Guinea. Each type has a different niche. This helps the pigeons avoid competition.

The Victoria crowned pigeon has a niche that includes eating fruits, berries, and large seeds. The pigeon nests in trees and searches for food on the ground of the forest.

The forests of New Guinea are home to Victoria crowned pigeons.

▷ **How do pigeons in the forests of New Guinea avoid competition?**

Why It Matters

As a living thing, you have a niche in your ecosystem. Your niche includes the roles you have at home and at school. How is your niche different from the niches of other people and animals in the ecosystem?

e-Journal Visit our Web site www.science.mmhschool.com to do a research project on competition among living things.

Think and Write

1. What is competition?

2. What things do organisms compete for?

3. What is a niche?

4. Why do some ecosystems have more types of organisms than others?

5. Critical Thinking A gazelle and a zebra both live in the same habitat and eat plants. How do they avoid competition?

L·I·N·K·S

MATH LINK

Solve a problem. Look at the chart on p. B43. What is the difference between the numbers of different types of trees in the temperate forest and the rain forest? Which forest has more types of trees?

WRITING LINK

Expository Writing Observe an animal every day for one week. How is it fit for its environment? How does it compete with other living things for resources? Use your observations to write a report about the animal's niche in the ecosystem.

LITERATURE LINK

Read *The Wolves' Winter* to learn how a pair of young wolves search for a new home. When you finish reading, think about how an animal in your neighborhood might search for food. Try the activities at the end of the book.

The Wolves' Winter

by Alice Pernick
illustrated by Diane Blasius

TECHNOLOGY LINK

LOG ON Visit www.science.mmhschool.com for more links.

Buzzing Bees and Flower Seeds

We all know that bees buzz from flower to flower. They are collecting nectar to turn into honey. But did you know that bees are also doing another important job? Bees are helping flowers to make new flowering plants.

Flowers attract bees so that bees will land on them and get covered with pollen.

When a bee lands on a flower and sips its nectar, the bee's hairy legs and body get covered with sticky flower pollen.

This sticky pollen is good for flowers. Flowers depend on bees to drop pollen on other flowers. That is called pollination, and it is how flowers reproduce.

Many flowers have both male and female parts. Pollen contains a flower's male cells. Once pollen lands on a new flower, it makes its way inside the flower to the female cells. There the male and female cells combine, and a seed is made. This seed can grow into a new flowering plant.

The more times a bee pollinates a flower, the more seeds the flower can make. That's good for the flower. It's also good for the bee. More flowers means more nectar. More nectar means more honey. More honey means more food for the bee—and enough honey for people, too!

Some flowers use brightly colored petals to lure bees. Others give off a sweet aroma that bees like. Bees take pollen from one flower to another.

Write About It

1. Bees and flowers both benefit from pollination. How?

2. What would happen to some flowers if bees suddenly disappeared?

LOG ON Visit **www.science.mmhschool.com** to learn more about bees.

Adaptations for Survival

Vocabulary

adaptation, B50
camouflage, B52
mimicry, B53

Get Ready

Could a goose perch on a tree branch? What kind of feet would be best for perching? Parts of animals are like tools. Each part has a job to do. The job of goose feet is to paddle through water. Although all birds have beaks, not all beaks are the same. How does a bird's beak help it stay alive?

Inquiry Skill

You experiment when you perform a test to support or disprove a hypothesis.

Explore Activity

How Does the Shape of a Bird's Beak Affect What It Eats?

Materials

chopsticks

spoon

clothespin

drinking straw

rubber worm

peanut in shell

rice

water in paper cup

Procedure: Design Your Own

1 Predict How does the shape of a bird's beak affect what it eats? Record your prediction.

2 Make a Model Look at the materials given to you. How will you use them? Record your plan.

3 Create a chart like the one here to record your data.

4 Follow your plan. Be sure to record all your observations.

Bird Beak Observations		
Type of Beak	Type of Food	Observations

Drawing Conclusions

1 Share your chart with your classmates. How are your results similar? How are they different? Why is it important to compare your results with those of your classmates?

2 Explain why different tools are better suited to different jobs.

3 Infer How does the shape of a bird's beak help it to eat the foods it needs? How do you know?

4 **FURTHER INQUIRY** **Experiment** Are different teeth better for eating different foods? How would you test your ideas?

Main Idea Living things have adaptations that help them survive.

What Is an Adaptation?

Tools work in different ways. Some tools are good for picking up small things. Other tools are better for picking up large things. Parts of organisms also work like tools. A bird uses its beak as a tool for eating. Different beak shapes are suited to different kinds of food.

The honeycreeper is a kind of bird. There are different types of honeycreepers. Each type has a beak that is shaped differently. Each beak shape is an **adaptation** (ad·uhp·TAY·shuhn). An adaptation is a special characteristic that helps an organism survive. How do different kinds of beaks help honeycreepers survive?

Honeycreeper beaks are just one example of adaptation. There are many others. In fact, most organisms have a variety of adaptations. Each adaptation helps the organism survive.

How is the wool of a lamb an adaptation? It keeps the lamb warm. A warm coat helps the lamb survive cold winter days. A giraffe's long neck is an adaptation, too. It helps the giraffe find food in high places. Finding food that others can't reach increases the giraffe's chances of survival.

Honeycreepers

Each type of honeycreeper has one of these three basic beak shapes.

A long, curved beak is good for eating nectar from flowers.

A beak that is short, thick, and strong is just right for eating seeds and nuts.

A straight beak is good for eating insects.

The bright coloring of a flower is an adaptation.

A frog has a long, sticky tongue and powerful legs. Both of these adaptations help the frog catch insects for food. The bright coloring of a flower is an adaptation. It attracts insects that help the flower reproduce. By reproducing, this type of plant survives.

READING **Compare and Contrast**

How are the different types of beaks on honeycreepers alike? How are they different?

The European Grass Snake can pretend it is dead. How does this adaptation help it survive?

A frog's sticky tongue and powerful legs are adaptations.

How Can Adaptations Protect Living Things?

Not all adaptations are important for getting food or reproducing. Many adaptations help protect an organism from harm.

For example, many animals have body colors or shapes that match their surroundings. Look at the animals on this page. When these animals stay still, a predator may not see them. These are examples of **camouflage** (KAM·uh·flahzh). Camouflage is an adaptation that allows an organism to blend in with its environment. It increases an animal's chance for survival. A white rabbit blends in with the snow, while brown rabbits match their forest habitat. In some cases body coloring can also hide a predator from its prey. The stripes of a tiger allow it to be unseen as it stands in tall grass.

Can you point out the pipefish among the plants? This fish holds its body straight up and down when it swims.

This Indian leaf butterfly is protected by camouflage. With its wings folded up, it looks like a dead leaf. Can you point out the butterfly?

Some animals are protected because they look like other animals. Looking like another organism is called **mimicry** (MIM·i·kree). Sometimes a gentle animal may look like a predator. For example, a snake and a snake mimic caterpillar look very much alike. One animal is a dangerous predator. The other is a harmless plant eater. Animals that eat caterpillars may avoid this one because it looks like a snake.

The wings of the grey butterfly have spots that look like eyes. These eyespots look like the eyes of a powerful hunter, the owl. A small bird looking for a meal would probably fly on by the butterfly.

The snake mimic caterpillar avoids being eaten by looking like a real snake.

▶ **How can color protect some living things?**

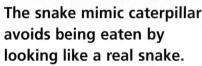

If you saw this butterfly among the leaves of a tree, what would you think you were looking at?

What Do Animals Do to Defend Themselves?

Animals have many ways of defending themselves against danger. Some animals have a keen sense of smell or hearing. These senses warn them of a nearby predator. Using these senses, the animals can hide or escape.

Some animals defend themselves by fighting with strong claws, sharp teeth, and powerful jaws. Others have special ways of chasing away a predator. These defenses are all important adaptations.

Some defenses are learned by animals. Other defenses are done by instinct. Instinct is a way of acting that an animal is born with. It does not need to be learned. The pictures here show instinct.

▷ **How do animals defend themselves?**

The porcupine fish inflates itself into a prickly ball when it senses a predator.

When a pill bug needs to defend itself, it curls up.

When the frill-necked lizard of Australia is attacked, it raises the stiff skin around its neck. It hisses and lashes its tail. It scares off even large snakes.

How Do Animals in Different Environments Adapt?

Organisms have adaptations that help them survive in their environment. Deserts are hot, dry places. There is little rain. Water is often deep underground. Wolves that live in a hot desert have thin coats. Cacti that live in a desert have thick stems to store water.

The tundra is a cold, dry place. Here the soil is frozen several inches down. Only very small trees are able to survive in this environment. Wolves that live in the tundra have thick fur.

▷ **How are wolves that live in the cold forest different from those that live in the hot desert?**

Adaptations in Different Environments

	Trees	Bears	Birds
Arctic tundra	Only very small trees grow in the tundra.	A polar bear has thick white fur.	A snowy owl has white. feathers
Desert	A mesquite tree has deep roots.	No bears live in the desert.	A roadrunner has brown feathers.

READING Charts

How are the colors of the animals alike? How are they different?

Design an Animal

You know that camouflage is one way animals keep safe. In this activity you will observe an area of your classroom. When you observe something, you use one or more of your senses to learn about the objects. You will use your observations of your classroom to help you design an animal that could hide in that environment.

Materials

construction paper

crayons

cotton balls

yarn

scissors

tape

Procedure

BE CAREFUL! Be careful when using scissors.

1 **Observe** Select an area to observe. This area is the environment for the organism that you will design. What do you notice about the area? What colors do you see? What textures do you feel? Record your observations.

2 Create a plan with a classmate. Make a list of features that would help an organism hide in this environment.

3 Use the materials given to you to create a plant or an animal that will blend into its surroundings. Put your plant or animal into its environment.

Drawing Conclusions

1 Describe the characteristics of the organism that you made. Explain why you included each one.

2 **Infer** Some animals can change the color of their body covering. Why might they do this?

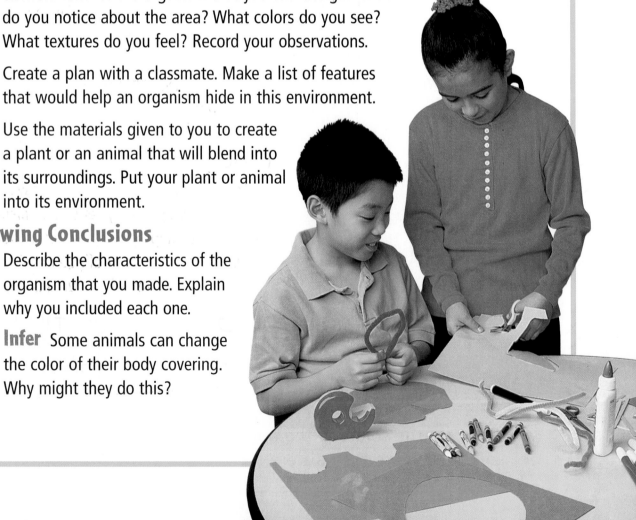

Why It Matters

People also have special adaptations that suit their environment. Your hands are a special adaptation. For example, your hands let you paint a picture, throw a ball, or play the piano. What other adaptations do you have?

e-Journal Visit our Web site www.science.mmhschool.com to do a research project on adaptations.

Think and Write

1. What is an adaptation?

2. What is camouflage? Give an example of camouflage.

3. Describe some adaptations that protect organisms.

4. INQUIRY SKILL Observe Look at the picture of the frog on page B51. What adaptations help it survive?

5. Critical Thinking Compare reptiles that live in the desert with those that live in the rain forest. In what ways would you expect them to be different? How might they be the same?

L·I·N·K·S

MATH LINK

Solve a problem. What if there are 15 honeycreeper families that live in a forest? If each family has five young birds, how many honeycreepers would live in this forest?

WRITING LINK

Writing That Compares Describe your favorite animal. Where does it live? What adaptations does it have to help it survive? Compare the adaptations with those of another animal, such as a horned toad or a chameleon. Write how the animals are alike and how they are different.

ART LINK

Create a poster. Design an ecosystem that shows several types of organisms using the adaptations that allow them to survive (such as camouflage or mimicry). Share your poster with your classmates.

TECHNOLOGY LINK

LOG ON Visit www.science.mmhschool.com for more links.

Changing Ecosystems

Get Ready

For many years Mount St. Helens was a sleeping volcano. Bears and elk roamed its forests. Fish swam in its streams. Wildflowers bloomed on its slopes. Then on May 18, 1980, Mount St. Helens erupted. This changed local habitats forever. How did these changes affect wildlife in the area?

Inquiry Skill

You predict when you state possible results of an event or an experiment.

Explore Activity

What Happens When Ecosystems Change?

Materials

3 predator cards:
  red hawk
  blue owl
  green snake

12 prey cards:
  4 red
  4 blue
  4 green

Procedure

1. Make the 3 predator cards and 12 prey cards listed in the Materials. Give each player one predator card. Stack the prey cards in the center of the table.

2. Take turns drawing a prey card. Keep only the prey cards that match the color of your predator card. Return all others. Play until one predator gets all four matching prey cards.

3. **Experiment** Add a card that says "fire" to the prey cards. Play the game again. Any predator who draws the fire card must leave the game. Return the fire card to the deck. Continue to play until a predator gets all four prey cards or all players are out.

Drawing Conclusions

1. What happened each time you played the game?

2. The fire card represented an ecosystem change. What effect did it have?

3. **Infer** What may happen when an ecosystem changes?

4. **FURTHER INQUIRY** **Predict** What might happen if you changed the number of prey cards? Try it.

Main Idea Changes in ecosystems affect the plants and animals that live there.

What Happens When Ecosystems Change?

Ecosystems can change. The change can be large or small. When a large change occurs, the organisms that live in that ecosystem are affected. Some have trouble surviving.

The eruption of Mount St. Helens in Washington State caused huge changes. A wind of hot steam and rock blasted the area. The wind lifted trees right out of the ground. Organisms that made their homes on or near trees were affected. The forest was buried in ash. When it rained, the ash turned muddy. Then the ash hardened into a tough crust that killed the plants underneath it. Other organisms near the ground had their habitats destroyed.

Animals that roamed the forest now had no homes. Some moved far from Mount St. Helens in search of new homes. Others could not find new habitats and died.

Melting snow and storms led to flooding along the Missouri River.

Spirit Lake below Mount St. Helens before the eruption

Spirit Lake after the eruption

Over time some organisms found new habitats. The fireweed plant began to grow right through the cracks in the crust. As one organism moved back in, others followed. After a few years, a new ecosystem began to form on the crust of ash. It is different from the ecosystem that was there before.

A volcanic eruption is one event that can change an ecosystem. A flood or drought can change an ecosystem, too. Floods are caused by heavy rains or snow. Rivers rise up over their banks and cover dry land. This drowns plants and changes habitats. A drought is the opposite of a flood. During a drought it doesn't rain for weeks or months. Rivers and lakes dry up. Plants and animals are affected. Other natural disasters, such as fires, earthquakes, and storms, can also change an ecosystem.

People change the ecosystem, too. People cut down trees and use the wood to build homes. They can make a wetland for a new supply of clean water.

▷ What are three things that can cause changes in habitats?

How do the conditions of Spirit Lake today determine how well these plants grow?

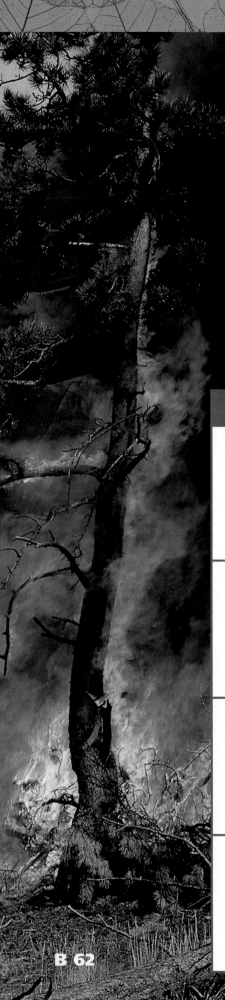

How Do Ecosystems Come Back?

After a big change, ecosystems usually come back. A fire can destroy almost all the habitats in a forest. How does a forest return? There are several stages the forest must go through.

READING Charts

1. What happens to a forest right after it is destroyed?

2. Why do animals return after the grasses and insects instead of before them?

How a Forest Comes Back After a Fire

Stage 1 Habitat destruction

Bulbs and seeds may survive underground. They begin to grow in the ash.

Stage 2 Grasses

Over time grasses cover the bare ground. The grasses add nutrients to the soil. They also provide a home for insects. The insects attract larger animals.

Stage 3 Larger plants

Small trees begin to grow. The trees block the sunlight. Without light the grasses begin to die.

Stage 4 Forest

Small trees are replaced by larger trees. The forest is the final stage.

Organisms respond to change in one of three ways. Some organisms respond to a change in their habitat by adjusting. The fireweed on Mount St. Helens was covered by crust. It adjusted to its new habitat by growing through the crust.

This box turtle survived a forest fire.

Some organisms **perish** (PER·ish). Organisms that perish do not survive. A mouse may survive a fire. Where will it find food after the fire? If it cannot meet its needs, it may not survive. Some organisms **relocate** (ree·LOH·kayt). An organism that relocates finds a new home. Trees were destroyed on Mount St. Helens. Birds that lived in the trees could fly to new trees.

▷ **What are two ways organisms can respond to change?**

Many animals are driven away as their habitats become towns. This moose is feeding near a house.

Are Living Things Dying Out?

Natural disasters such as fires and floods can destroy habitats. People can harm habitats, too. When people replace plains or forests with new towns, organisms may lose their homes or become overcrowded. Pollution can harm organisms, too. Some animals are hunted by people.

All these things are threats to plants and animals. In many cases organisms become **endangered** (en·DAYN·juhrd). An organism is called endangered when there are very few of its kind left.

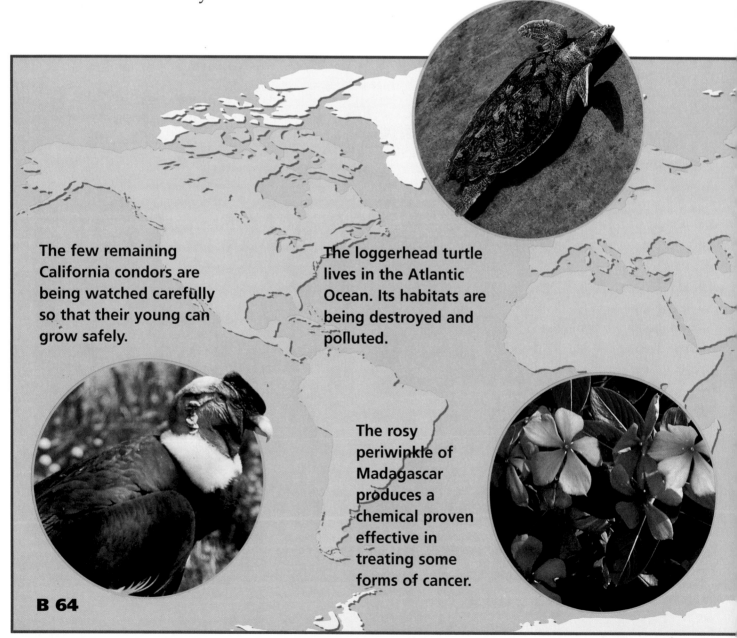

The few remaining California condors are being watched carefully so that their young can grow safely.

The loggerhead turtle lives in the Atlantic Ocean. Its habitats are being destroyed and polluted.

The rosy periwinkle of Madagascar produces a chemical proven effective in treating some forms of cancer.

People are finding ways to protect many endangered organisms. Some endangered animals live in national parks, where they are watched by rangers. Laws are being passed to stop the hunting of endangered animals.

▶ **When does an animal become endangered?**

The Bengal tiger lives in southern Asia. Hunters have killed almost all of this kind of tiger.

QUICK LAB

Crowd Control

FOLDABLES™ Make a Two-Tab Book. (See p. R41.) Label the book as shown.

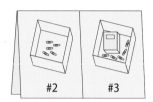

#2   #3

1. Toss 20 paper clips in a small box. Remove any two that touch each other.

2. Gather the paper clips that are left in the box. Toss them again and remove any two that touch. Repeat until there are no clips left. Count how many tosses you made. Record your answer on the first tab of your Foldables book.

3. Repeat steps 1 and 2. This time, put a book in the box so there is less room for the clips to move in. Record the number of tosses you made on the second tab.

4. **Infer** When organisms are crowded together, how do their chances of survival change? Write your answer on the inside of the Foldables book.

Have Living Things Died Out?

The answer is "Yes." For example, all the dinosaurs that lived millions of years ago became **extinct** (ek·STINGKT). *Extinct* means that there are no more of that type of organism alive. Extinct animals include dinosaurs. Scientists are not sure what caused these animals to die out. However, we do know why many other organisms became extinct. Changes in habitats, pollution, and hunting have killed off many living things.

Many organisms are becoming extinct even today, as rain forests are cleared. In the last 500 years, 500 kinds of organisms have become extinct in what is now the United States. Endangered organisms that you read about may become extinct if we do not protect them.

READING **Compare and Contrast** What are some of the different ways organisms can become extinct?

The saber-toothed cat was similar to today's tiger. It became extinct thousands of years ago, when its prey became extinct.

Dodo birds once covered the island of Mauritius near Africa. They were hunted by people until they became extinct in 1680.

L·I·N·K·S

Why It Matters

Why should you care about endangered organisms? When an organism becomes extinct, it is gone forever. This changes the food web it belonged to. It also may affect products we get from the organism.

e-Journal Visit our Web site www.science.mmhschool.com to do a research project on endangered organisms.

Think and Write

1. What causes ecosystems to change?

2. How might a forest recover after a fire or flood?

3. How do organisms respond when their habitats change?

4. What happened when there were no laws to control hunting?

5. **Critical Thinking** A forest is cut down to build a parking lot. Why is this habitat change more serious than a forest fire?

MATH LINK

Make a graph. Look out a window for ten minutes. Record the number of animals you see. Repeat your observations each day at the same time at the same window for one week. Draw a bar graph to show how the number of animals changes each day. Use writing to explain your graph.

WRITING LINK

Persuasive Writing Describe an endangered organism that you want to save. Write an article about the organism for a newspaper. Tell why it is important to save this endangered organism.

LITERATURE LINK

Read *Rescue at First Encounter Beach* to learn how a pod of pilot whales becomes beached. When you finish reading, write a list of the ways the rescuers helped the whales. Try the activities at the end of the book.

Rescue at First Encounter Beach
by Pearl Neuman
Illustrated by Anita Stewart

TECHNOLOGY LINK

LOG ON Visit www.science.mmhschool.com for more links.

Chapter 4 Review

Vocabulary

Fill each blank with the best word from the list.

adaptation, B50   **mimicry,** B53

camouflage, B52   **niche,** B44

competition, B42   **perish,** B63

endangered, B64   **relocate,** B63

extinct, B66

1. Organisms that all want the same thing are in _____.

2. The job or role an organism has in an ecosystem is its _____.

3. A characteristic that helps an organism survive in its environment is a(n) _____.

4. The white fur of the polar bear is an example of _____.

5. An adaptation in which one organism imitates or looks like another is called _____.

6. An organism that is _____ is gone forever.

When an ecosystem changes, living things may   **7.** or **8.** .
When living things do not survive, that type of organism may become   **9.** or **10.** .

Test Prep

11. Predators often compete for the same _____.

    A adaptation

    B prey

    C habitat

    D ecosystem

12. All the following are adaptations EXCEPT _____.

    F camouflage

    G mimicry

    H the shape of a bird's beak

    J the prey in a habitat

13. A snowy owl most likely lives in the _____.

    A tundra

    B desert

    C rain forest

    D temperate forest

14. An ecosystem can be changed by _____.

    F floods

    G droughts

    H fires

    J all of the above

15. Which of the following animals is now extinct?

A dodo bird

B pigeon

C blue whale

D elephant

Concepts and Skills

16. Reading in Science Look at the pictures below. Compare and contrast the two pictures of Spirit Lake.

Before   **After**

17. Critical Thinking One group of rabbits lives on an island without predators. A second group of rabbits lives on a different island with predators. Which group is more likely to be faster runners? Explain.

18. Safety You are exploring changes in a habitat in a nearby park. Write down some rules you might follow to be sure your study is safe.

19. Scientific Methods What animals (not pets) in your neighborhood use camouflage? Plan to make a survey of animals in your neighborhood to find out. Tell what characteristics allow each animal to blend in with its surroundings.

20. INQUIRY SKILL **Observe** Draw a picture of your favorite animal. What adaptations does your animal have? What purpose does each adaptation serve?

Did You Ever Wonder?

INQUIRY SKILL **Form a Hypothesis** You have seen that many animals have unusual behaviors that help them to survive. Think of a question such as, why does a spider not get caught in its own web? How can you find the answer?

 LOG ON Visit **www.science.mmhschool.com** to boost your test scores.

Meet a Scientist

Dr. Francisco Dallmeier

WILDLIFE CONSERVATIONIST

Dr. Dallmeier and a giant sea turtle

In the rain forest of Gabon, in Africa, elephants, leopards, and reptiles prowl the land. Francisco Dallmeier and a team of scientists wanted to identify every type of plant and animal in the area. Gabon is a rain forest near the Congo River. "For the future of Gabon, it's important to know what we have there."

The team tracked down and recorded 159 types of reptiles and amphibians. They found 70 kinds of fish, 140 types of trees, and dozens of mammal species. Dallmeier also discovered several plant and animal species that had never been seen before.

Dallmeier's work has helped people understand the importance of the Gabon rain forest. He says, "We need to protect these plant and animal species. They are valuable to the natural balance of the area and our Earth."

LOG ON Visit www.science.mmhschool.com to learn more about rain forests.

TIME FOR KIDS®

TOP 5 Rain Forest Facts

1. About 2,000 trees per minute are cut down in the rain forests.
2. Rain forests are home to millions of plant and animal species.
3. On one day in 1987, a total of 7,603 fires were burning in the Amazon rain forest.
4. Four square miles of rain forest can contain as many as 1,500 species of flowering plants.
5. Most of the plants useful in cancer treatment are found only in the rain forests.

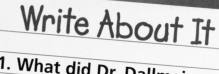

Write About It

1. What did Dr. Dallmeier find in the Gabon rain forest?
2. Why does Dr. Dallmeier want to help protect plants and animals in the Gabon rain forest?

Composting Recycles

Your goal is to find out if a compost pile will heat up.

Materials

2 2L clear plastic drink bottles	grass clippings, leaves,
marker	plant peelings,
scissors	newspaper strips
clear 1-inch tape	thermometer
1 cup of soil	water

What to Do

1. Have an adult help you make the compost container. Use the diagram.

2. Fill the container with grass clippings, leaves, newspaper strips, plant peelings, or a combination of these materials. Add the soil and mix it with the other materials. Moisten the substances in the container. Cover the container with the top of the first bottle (A).

3. Observe the container each day. Record what you observe on a chart.

4. With an adult, open the top and place the thermometer in the container twice each week.

5. Record the temperature on a chart.

Draw Conclusions

Did the compost pile heat up? Write a paragraph about your findings. Explain how composting can be used to recycle discarded plant and animal material.

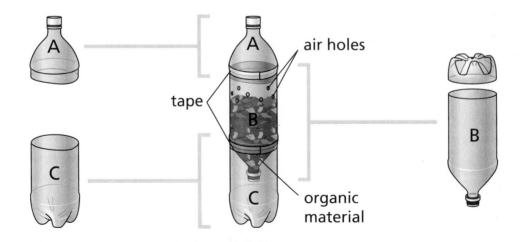

For Your Reference

Science Handbook

Health Handbook

Glossary

Index

Units of Measurement

Temperature

1. The temperature is 77 degrees Fahrenheit.

2. That is the same as 25 degrees Celsius.

3. Water boils at 212 degrees Fahrenheit.

4. Water freezes at 0 degrees Celsius.

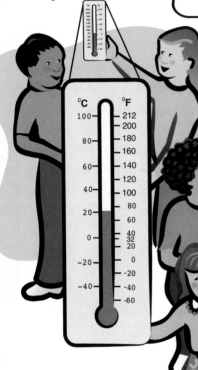

Length and Area

1. This classroom is 10 meters wide and 20 meters long.

2. That means the area is 200 square meters.

Mass and Weight

1. That baseball bat weighs 32 ounces.

2. 32 ounces is the same as 2 pounds.

3. The mass of the bat is 907 grams.

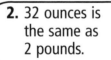

Measurement

Volume of Fluids

Rate

Weight/Force

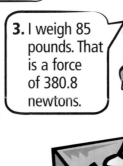

1. This bottle of juice has a volume of 1 liter.

2. That is a little more than 1 quart.

3. I weigh 85 pounds. That is a force of 380.8 newtons.

1. She can walk 20 meters in 5 seconds.

2. That means her speed is 4 meters per second.

Table of Measurements

SI (International System) of Units	English System of Units
Temperature Water freezes at 0 degrees Celsius (°C) and boils at 100°C.	**Temperature** Water freezes at 32 degrees Fahrenheit (°F) and boils at 212°F.
Length and Distance 10 millimeters (mm) = 1 centimeter (cm) 100 centimeters = 1 meter (m) 1,000 meters = 1 kilometer (km)	**Length and Distance** 12 inches (in.) = 1 foot (ft) 3 feet = 1 yard (yd) 5,280 feet = 1 mile (mi)
Volume 1 cubic centimeter (cm^3) = 1 milliliter (mL) 1,000 milliliters = 1 liter (L)	**Volume of Fluids** 8 fluid ounces (fl oz) = 1 cup (c) 2 cups = 1 pint (pt) 2 pints = 1 quart (qt) 4 quarts = 1 gallon (gal)
Mass 1,000 milligrams (mg) = 1 gram (g) 1,000 grams = 1 kilogram (kg)	**Weight** 16 ounces (oz) = 1 pound (lb) 2,000 pounds = 1 ton (T)
Area 1 square kilometer (km^2) = 1 km x 1 km 1 hectare = 10,000 square meters (m^2)	**Rate** mph = miles per hour
Rate m/s = meters per second km/h = kilometers per hour	
Force 1 newton (N) = 1 kg x 1m/s^2	

Use a Hand Lens

You use a hand lens to magnify an object, or make the object look larger. With a hand lens, you can see details that would be hard to see without the hand lens.

Magnify a Piece of Cereal

1. Place a piece of your favorite cereal on a flat surface. Look at the cereal carefully. Draw a picture of it.
2. Hold the hand lens so that it is just above the cereal. Look through the lens, and slowly move it away from the cereal. The cereal will look larger.

3. Keep moving the hand lens until the cereal begins to look blurry. Then move the lens a little closer to the cereal until you can see it clearly.
4. Draw a picture of the cereal as you see it through the hand lens. Fill in details that you did not see before.
5. Repeat this activity using objects you are studying in science. It might be a rock, some soil, a seed, or something else.

Collect Data

Use a Microscope

Hand lenses make objects look several times larger. A microscope, however, can magnify an object to look hundreds of times larger.

Examine Salt Grains

1. Place the microscope on a flat surface. Always carry a microscope with both hands. Hold the arm with one hand, and put your other hand beneath the base.
2. Look at the drawing to learn the different parts of the microscope.
3. Move the mirror so that it reflects light up toward the stage. Never point the mirror directly at the Sun or a bright light. Bright light can cause permanent eye damage.
4. Place a few grains of salt on the slide. Put the slide under the stage clips on the stage. Be sure that the salt grains are over the hole in the stage.
5. Look through the eyepiece. Turn the focusing knob slowly until the salt grains come into focus.
6. Draw what the grains look like through the microscope.
7. Look at other objects through the microscope. Try a piece of leaf, a strand of human hair, or a pencil mark.
8. Draw what each object looks like through the microscope. Do any of the objects look alike? If so, how? Are any of the objects alive? How do you know?

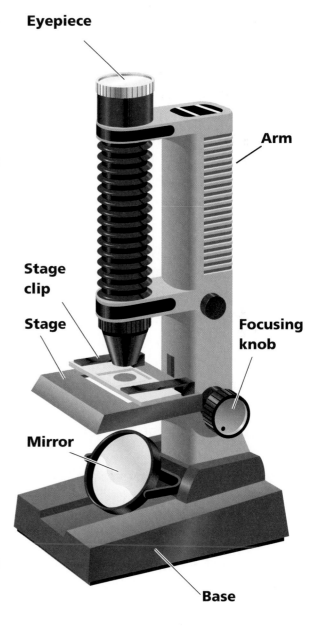

Eyepiece

Arm

Stage clip

Stage

Focusing knob

Mirror

Base

Measure Time

You use timing devices to measure how long something takes to happen. Some timing devices you use in science are a clock with a second hand and a stopwatch. Which one is more accurate?

Comparing a Clock and a Stopwatch

1. Look at a clock with a second hand. The second hand is the hand that you can see moving. It measures seconds.

2. Get an egg timer with falling sand. When the second hand of the clock points to 12, tell your partner to start the egg timer. Watch the clock while the sand in the egg timer is falling.

3. When the sand stops falling, count how many seconds it took. Record this measurement. Repeat the activity, and compare the two measurements.

4. Look at a stopwatch. Click the button on the top right. This starts the time. Click the button again. This stops the time. Click the button on the top left. This sets the stopwatch back to zero. Notice that the stopwatch tells time in hours, minutes, seconds, and hundredths of a second.

5. Repeat the activity in steps 1–3, but use the stopwatch instead of a clock. Make sure the stopwatch is set to zero. Click the top right button to start timing. Click the

button again when the sand stops falling. Make sure you and your partner time the sand twice.

0 minutes   **25 seconds 72 hundredths of a second**

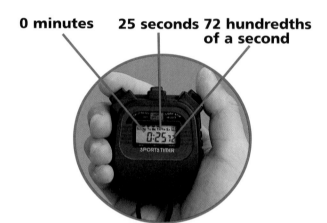

More About Time

1. Use the stopwatch to time how long it takes an ice cube to melt under cold running water. How long does an ice cube take to melt under warm running water?

2. Match each of these times with the action you think took that amount of time.

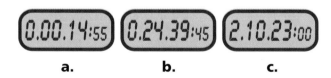

a.   b.   c.

1. A Little League baseball game
2. Saying the Pledge of Allegiance
3. Recess

Measure Length

You measure length to find out how long something is or how far away something is.

Find Length with a Ruler

1. Look at this section of a ruler. Each centimeter (cm) is divided into 10 millimeters (mm). How long is the paper clip?

2. The length of the paper clip is 3 centimeters plus 2 millimeters. You can write this length as 3.2 centimeters.

3. Place a ruler on your desk. Lay a pencil against the ruler so that one end of the pencil lines up with the left edge of the ruler. Record the length of the pencil.

4. Measure the length of another object. What unit of measure did you use?

5. Ask a partner to measure the same object. Compare your answers. Explain how measurements can be slightly different even if the item measured is the same.

Measuring Area

Area is the amount of surface something covers. To find the area of a rectangle, multiply the rectangle's length by its width. For example, the rectangle here is 3 centimeters long and 2 centimeters wide. Its area is 3 cm x 2 cm = 6 square centimeters. You write the area as 6 cm².

1. Find the area of your science book. Measure the book's length to the nearest centimeter. Measure its width.

2. Multiply the book's length by its width. Remember to put the answer in cm².

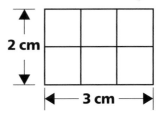

2 cm

← 3 cm →

More About Length

Another tool that measures length is called a caliper. It measures distances and thicknesses. It has two movable, curved legs on a hinge. Try measuring a baseball with a caliper. How wide is it from one side to the other?

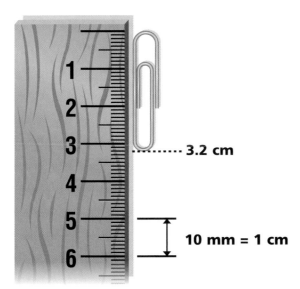

.... 3.2 cm

10 mm = 1 cm

Measure Mass

Mass is the amount of matter an object has. You use a balance to measure mass. To find the mass of an object, you balance it with objects whose masses you know.

Measure the Mass of a Box of Crayons

1. Place the balance on a flat, level surface.
2. The pointer should point to the middle mark. If it does not, move the slider a little to the right or left to balance the pans.
3. Gently place a box of crayons on the left pan. Add gram masses to the right pan until the pans are balanced.
4. Count the numbers on the masses that are in the right pan. The total is the mass of the box of crayons, in grams.

5. Record this number. After the number, write a *g* for "grams."

More About Mass

What would happen if you replaced the crayons with a paper clip or a pineapple? You may not have enough masses to balance the pineapple. It has a mass of about 1,000 grams. That's the same as 1 kilogram, because *kilo* means "1,000." Measure other objects and record your measurements.

Measure Volume

Have you ever used a measuring cup? Measuring cups measure the volume of liquids. Volume is the amount of space something takes up. In science you use special measuring cups called beakers and graduated cylinders. These containers are marked in milliliters (mL).

Measure the Volume of a Liquid

1. Look at the beaker and at the graduated cylinder. The beaker has marks for each 25 mL up to 200 mL. The graduated cylinder has marks for each 1 mL up to 100 mL.

2. The surface of the water in the graduated cylinder curves up at the sides. You measure the volume by reading the height of the water at the flat part. What is the volume of water in the graduated cylinder? How much water is in the beaker?

3. Pour 50 mL of water from a pitcher into a graduated cylinder. The water should be at the 50-mL mark on the graduated cylinder. If you go over the mark, pour a little water back into the pitcher.

4. Pour the 50 mL of water into a beaker.

5. Repeat steps 3 and 4 using 30 mL, 45 mL, and 25 mL of water.

6. Measure the volume of water you have in the beaker. Do you have about the same amount of water as your classmates?

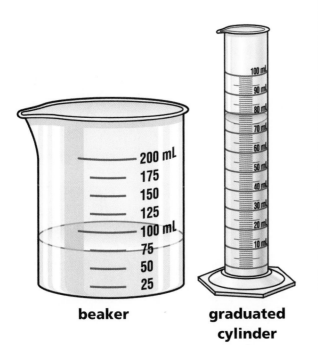

beaker   graduated cylinder

Measure Weight/Force

You use a spring scale to measure weight. An object has weight because the force of gravity pulls down on the object. Therefore, weight is a force. Like all forces, weight is measured in newtons (N).

Measure the Weight of an Object

1. Look at your spring scale to see how many newtons it measures. See how the measurements are divided. The spring scale shown here measures up to 10 N. It has a mark for every 1 N.
2. Hold the spring scale by the top loop. Put the object to be measured on the bottom hook. If the object will not stay on the hook, place it in a net bag. Then hang the bag from the hook.
3. Let go of the object slowly. It will pull down on a spring inside the scale. The spring is connected to a pointer. The pointer on the spring scale shown here is a small arrow.

4. Wait for the pointer to stop moving. Read the number of newtons next to the pointer. This is the object's weight. The mug in the picture weighs 3 N.

More About Spring Scales

You probably weigh yourself by standing on a bathroom scale. This is a spring scale. The force of your body stretches a spring inside the scale. The dial on the scale is probably marked in pounds—the English unit of weight. One pound is equal to about 4.5 newtons.

Here are some spring scales you may have seen.

Measure Temperature

Temperature is how hot or cold something is. You use a thermometer to measure temperature. A thermometer is made of a thin tube with colored liquid inside. When the liquid gets warmer, it expands and moves up the tube. When the liquid gets cooler, it contracts and moves down the tube. You may have seen most temperatures measured in degrees Fahrenheit (°F). Scientists measure temperature in degrees Celsius (°C).

Read a Thermometer

1. Look at the thermometer shown here. It has two scales—a Fahrenheit scale and a Celsius scale. Every 20 degrees on each scale has a number.
2. What is the temperature shown on the thermometer? At what temperature does water freeze? Give your answers in °F and in °C.

How Is Temperature Measured?

1. Fill a large beaker about one-half full of cool water. Hold the thermometer in the water by using a clamp. Do not let the thermometer bulb touch the beaker.
2. Wait until the liquid in the tube stops moving—about a minute. Read and record the temperature. Record the temperature scale you used.

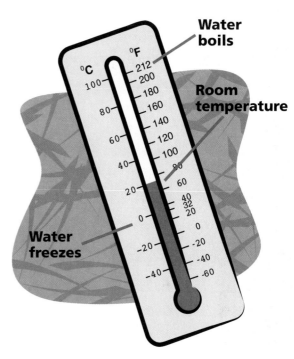

Water boils

Room temperature

Water freezes

3. Place the beaker with the thermometer on a hot plate and warm the beaker for two minutes. Be careful of the hot plate and warm water.
4. Record the temperature of the water. Use the same temperature scale you chose in Step 2.

Use Calculators: Add and Subtract

Sometimes after you make measurements, you have to add or subtract your numbers. A calculator helps you do this.

Add and Subtract Rainfall Amounts

The table shows the amount of rain that fell in a town each week during the summer.

Week	Rain (cm)
1	3
2	5
3	2
4	0
5	1
6	6
7	4
8	0
9	2
10	2
11	6
12	5

1. Make sure the calculator is on. Press the **ON** key.
2. To add the numbers, enter a number and press **+**. Repeat until you enter the last number. Then press **=**. You do not have to enter the zeros. Your total should be 36.

3. What if you found out that you made a mistake in your measurement? Week 1 should be 2 cm less, week 6 should be 3 cm less, week 11 should be 1 cm less, and week 12 should be 2 cm less. Subtract these numbers from your total. You should have 36 displayed on the calculator. Press **−**, and enter the first number you want to subtract. Repeat until you enter the last number. Then press **=**.

Use Technology

Use Calculators: Multiply and Divide

Sometimes after you make measurements, you have to multiply or divide your measurements to get other information. A calculator helps you multiply and divide, especially if the numbers have decimal points.

Multiply Decimals

What if you are measuring the width of your classroom? You discover that the floor is covered with tiles and the room is exactly 32 tiles wide. You measure a tile, and it is 22.7 centimeters wide. To find the width of the room, you can multiply 32 by 22.7.

1. Make sure the calculator is on. Press the **ON** key.
2. Press **3** and **2**.
3. Press **×**.
4. Press **2**, **2**, **·**, and **7**.
5. Press **=**. Your total should be 726.4. That is how wide the room is in centimeters.

Divide Decimals

Now what if you wanted to find out how many desks placed side by side would be needed to reach across the room? You measure one desk, and it is 60 centimeters wide. To find the number of desks needed, divide 726.4 by 60.

Remember that numbers have different values depending on what position they are in. A six in the ones place means six. In the tens place it means 60.

1. Turn the calculator on.
2. Press **7**, **2**, **6**, **·**, and **4**.
3. Press **÷**.
4. Press **6** and **0**.
5. Press **=**. Your total should be about 12.1. This means you can fit 12 desks across the room with a little space left over.

Suppose the room was 35 tiles wide. How wide would the room be? How many desks would fit across it?

Use Computers

A computer has many uses. The Internet connects your computer to many other computers around the world, so you can collect all kinds of information. You can use a computer to show this information and write reports. Best of all, you can use a computer to explore, discover, and learn.

You can also get information from CD-ROMs. They are computer disks that can hold large amounts of information. You can fit a whole encyclopedia on one CD-ROM.

Use Computers for a Project

Here's a project that uses computers. You can do the project in a group.

1. Use a collecting net to gather a soil sample from a brook or stream. Collect pebbles, sand, and small rocks. Keep any small plants also. Return any fish or other animals to the stream right away.

2. After the sample has dried, separate the items in the sample. Use a camera to photograph the soil, pebbles, small rocks, and plants.

3. Each group can use one of the photos to help them start their research. Try to find out what type of rocks or soil you collected.

4. Use the Internet for your research. Find a map and mark your area on it. Identify the type of soil. What types of plants grow well in that type of soil?

5. Find Web sites from an agency such as the Department of Environmental Protection. Contact the group. Ask questions about samples you collected.

6. Use CD-ROMS or other sources from the library to find out how the rocks and soil in your sample formed.

Use Technology

7. Keep the information you have gathered in a folder. Review it with your group and use it to write a group report about your soil sample.

8. Each group will present and read a different part of the report. Have an adult help you to record your reports on a video recorder. Show your photographs in the video and explain what each represents. If you'd like, use music or other sounds to accompany the voices on the video recorder.

9. Make a list of computer resources you used to make your report. List Web sites, CD-ROM titles, or other computer resources. Show or read the list at the end of your presentation.

10. Discuss how the computer helped each group to do their report. What problems did each group encounter using the computer? How were the problems solved?

Make Graphs to Organize Data

Graphs can help organize data. Graphs make it easy to spot trends and patterns. There are many kinds of graphs.

Bar Graphs

A bar graph uses bars to show information. For example, what if you are growing a plant? Every week you measure how high the plant has grown. Here is what you find.

Week	Height (cm)
1	1
2	3
3	6
4	10
5	17
6	20
7	22
8	23

The bar graph at the right organizes the measurements so you can easily compare them.

1. Look at the bar for Week 2. Put your finger at the top of the bar. Move your finger straight over to the left to find how many centimeters the plant grew by the end of Week 2.
2. Between which two weeks did the plant grow most?
3. Look at the 0 on the graph. Is it just a label on a scale or does it have a meaning in the graph? Explain.

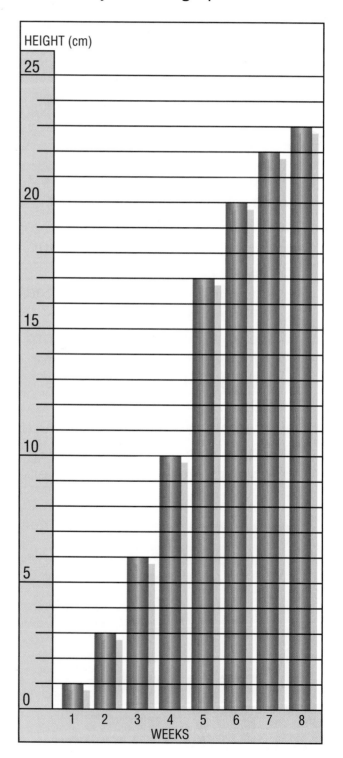

Represent Data

Pictographs

A pictograph uses symbols, or pictures, to show information. What if you collect information about how much water your family uses each day? Here is what you find.

Activity	Water Used Each Day (L)
Drinking	10
Showering	100
Bathing	120
Brushing teeth	40
Washing dishes	80
Washing hands	30
Washing clothes	160
Flushing toilet	50

You can organize this information into the pictograph shown here. In this pictograph each bottle means 20 liters of water. A half bottle means half of 20, or 10 liters of water.

1. Which activity uses the most water?
2. Which activity uses the least water?

Line Graphs

A line graph shows how information changes over time. What if you measure the temperature outdoors every hour starting at 6 A.M.? Here is what you find.

Time	Temperature (°C)
6 A.M.	10
7 A.M.	12
8 A.M.	14
9 A.M.	16
10 A.M.	18
11 A.M.	20

Now collect outside temperatures on your own each hour. Follow these steps to make a line graph.

1. Make a scale along the bottom and side of the graph as shown. Label the scales.
2. Plot points on the graph.
3. Connect the points with a line.
4. How do the temperatures and times relate to each other? Compare your graph to the one shown.

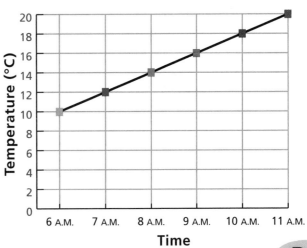

Make Maps, Tables, Charts

Locate Places

A map is a drawing that shows an area from above. Most maps have numbers and letters along the top and side. What if you wanted to find the library on the map below? It is located at D7. Place a finger on the letter D along the side of the map and another finger on the number 7 at the top. Then move your fingers straight across and down the map until they meet. The library is located where D and 7 meet.

1. What building is located at G3?
2. The hospital is located three blocks south and three blocks east of the library. What is its number and letter?
3. Make a map of an area in your community. It might be a park or the area between your home and school. Include numbers and letters along the top and side. Use a compass to find north, and mark north on your map. Exchange maps with classmates.

Idea Maps

The map below left shows how places are connected to each other. Idea maps, on the other hand, show how ideas are connected to each other. Idea maps help you organize information about a topic.

Look at the idea map below. It connects ideas about water. This map shows that Earth's water is either fresh water or salt water. The map also shows four sources of fresh water. You can see that there is no connection between "rivers" and "salt water" on the map. This reminds you that salt water does not flow in rivers.

Make an idea map about a topic you are learning in science. Your map can include words, phrases, or even sentences. Arrange your map in a way that makes sense to you and helps you understand the ideas.

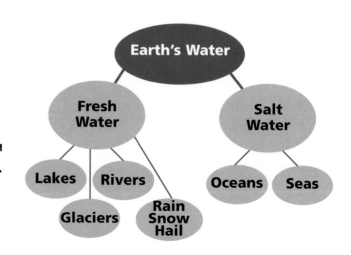

Make Tables and Charts to Organize Data

Tables help to organize data during experiments. Most tables have columns that run up and down, and rows that run across. The columns and rows have headings that tell you what kind of data goes in each part of the table.

A Sample Table

What if you are going to do an experiment to find out how long different kinds of seeds take to sprout? Before you begin the experiment, you should set up your table. Follow these steps.

1. In this experiment you will plant 20 radish seeds, 20 bean seeds, and 20 corn seeds. Your table must show how many of each kind of seed sprouted on days 1, 2, 3, 4, and 5.

2. Make your table with columns, rows, and headings. You might use a computer. Some computer programs let you build a table with just the click of a mouse. You can delete or add columns and rows if you need to.

3. Give your table a title. Your table could look like the one here.

Make a Table

Plant 20 bean seeds in each of two trays. Keep each tray at a different temperature and observe the trays for seven days. Make a table to record, examine, and evaluate the information of this experiment. How do the columns, rows, and headings of your table relate to one another?

Make a Chart

A chart is simply a table with pictures, as well as words to label the rows or columns. Make a chart that shows the information of the above experiment.

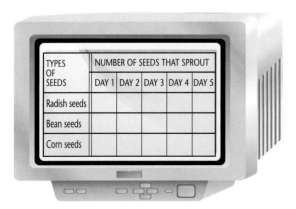

TYPES OF SEEDS	NUMBER OF SEEDS THAT SPROUT				
	DAY 1	DAY 2	DAY 3	DAY 4	DAY 5
Radish seeds					
Bean seeds					
Corn seeds					

The Skeletal System

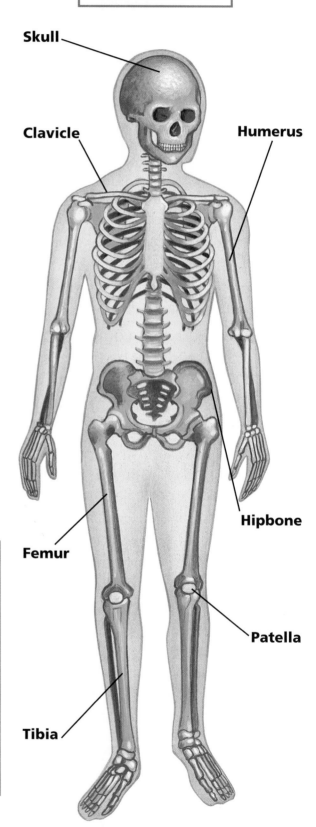

The Skeleton

The skeleton is a system of the human body. It is the frame that supports the body. The skeleton is made up of bones and has several jobs.

- It gives the body its shape.
- It protects organs in the body.
- It works with muscles to move the body.

Each of the 206 bones of the skeleton is the size and shape best fitted to do its job. For example, long and strong leg bones support the body's weight. The skull protects the brain. The hip bone helps you move.

1. What is the skeleton?
2. Describe several jobs of bones.

Labels on diagram: Skull, Clavicle, Humerus, Hipbone, Femur, Patella, Tibia

CARE!

- ● Exercise to keep your skeletal system in good shape.
- ● Don't overextend your joints.
- ● Eat foods rich in vitamins and minerals. Your bones need the minerals calcium and phosphorus to grow strong.

Bones

1 A bone is covered with a tough but thin membrane that has many small blood vessels. The blood vessels bring nutrients and oxygen to the living parts of the bone and remove wastes.

2 Inside some bones is a soft tissue known as marrow. Yellow marrow is made mostly of fat cells and is one of the body's energy reserves. It is usually found in the long, hollow spaces of long bones.

3 Part of the bone is compact, or solid. It is made up of living bone cells and non-living materials. The nonliving part is made up of layers of hardened minerals such as calcium and phosphorus. In between the mineral layers are living bone cells.

4 Red marrow fills the spaces in spongy bone. Red marrow makes new red blood cells, germ-fighting white blood cells, and cell fragments that stop a cut from bleeding.

5 Part of the bone is made of bone tissue that looks like a dry sponge. It is made of strong, hard tubes. It is also found in the middle of short, flat bones.

CARE!

- Eat foods rich in vitamins and minerals. Your bones need the minerals calcium and phosphorus to grow strong.

- Be careful! Avoid sprains and fractures.

- Get help in case of injury.

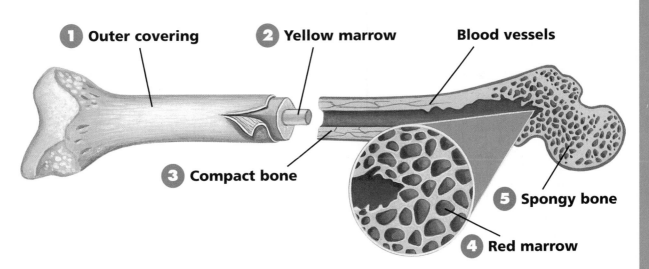

1 Outer covering   **2** Yellow marrow   Blood vessels

3 Compact bone

5 Spongy bone

4 Red marrow

R 21

Joints

The skeleton has different types of joints. A joint is a place where two or more bones meet. Joints can be classified into three major groups—immovable joints, partly movable joints, and movable joints.

Types of Joints

IMMOVABLE JOINTS

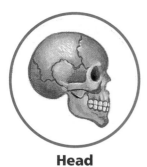

Head

Immovable joints are places where bones fit together too tightly to move. Nearly all the 29 bones in the skull meet at immovable joints. Only the lower jaw can move.

PARTLY MOVABLE JOINTS

Partly movable joints are places where bones can move only a little. Ribs are connected to the breastbone with these joints.

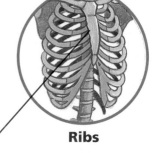

Breastbone

Ribs

MOVABLE JOINTS

Movable joints are places where bones can move easily. Use the information below to describe each type of movable joint. Explain how each type of joint allows movement.

Gliding joint

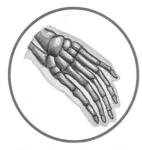

Hand and wrist

Small bones in the wrists and ankles meet at gliding joints. The bones can slide against one another. A gliding joint is similar to a sliding door. These joints allow some movement in all directions.

The hips are examples of ball-and-socket joints. The ball of one bone fits into the socket, or cup, of another bone. These joints allow bones to move back and forth, in a circle, and side to side.

Ball-and-socket joint

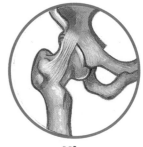

Hip

Hinge joint

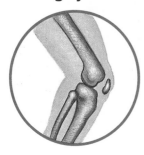

Knee

The knees are hinge joints. A hinge joint is similar to a door hinge. It allows bones to move back and forth in one direction.

The joint between the skull and neck is a pivot joint. It allows the head to move up and down, and side to side. A pivot joint is similar to a compass.

Pivot joint

Neck

The Muscular System

1 A message from your brain causes this muscle, called the biceps, to contract. When a muscle contracts, it becomes shorter and thicker. As the biceps contracts, it pulls on the arm bone it is attached to.

2 Most muscles work in pairs to move bones. This muscle, called the triceps, relaxes when the biceps contracts. When a muscle relaxes, it becomes longer and thinner.

3 To straighten your arm, a message from your brain causes the triceps to contract. When the triceps contracts, it pulls on the bone it is attached to.

4 As the triceps contracts, the biceps relaxes. Your arm straightens.

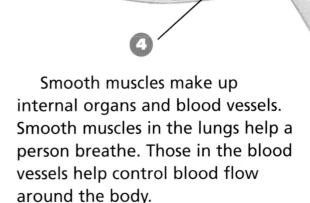

Three types of muscles make up the body—skeletal muscle, cardiac muscle, and smooth muscle.

The muscles that are attached to and move bones are called skeletal muscles. These muscles are attached to bones by a tough cord called a tendon. Skeletal muscles pull bones to move them. Muscles do not push bones.

Cardiac muscles are found in only one place in the body—the heart. The walls of the heart are made of strong cardiac muscles. When cardiac muscles contract, they squeeze blood out of the heart. When cardiac muscles relax, the heart fills with more blood.

Smooth muscles make up internal organs and blood vessels. Smooth muscles in the lungs help a person breathe. Those in the blood vessels help control blood flow around the body.

1. Name the three types of muscles.
2. Describe how muscles cause the body to move.

CARE!

- **Exercise to strengthen your muscles.**

- **Eat the right foods, and get plenty of rest.**

The Circulatory System

The circulatory system consists of the heart, blood vessels, and blood. Circulation is the flow of blood through the body. Blood is a liquid that contains red blood cells, white blood cells, and platelets. Red blood cells carry oxygen and nutrients to cells. White blood cells work to fight germs that enter the body. Platelets are cell fragments that make the blood clot.

The heart is a muscular organ about the size of a fist. It beats about 70 to 90 times a minute, pumping blood through the blood vessels. Arteries carry blood away from the heart. Some arteries carry blood to the lungs, where the cells pick up oxygen. Other arteries carry oxygen-rich blood from the lungs to all other parts of the body. Veins carry blood from other parts of the body back to the heart. Blood in most veins carries the wastes released by cells and has little oxygen. Blood flows from arteries to veins through narrow vessels called capillaries.

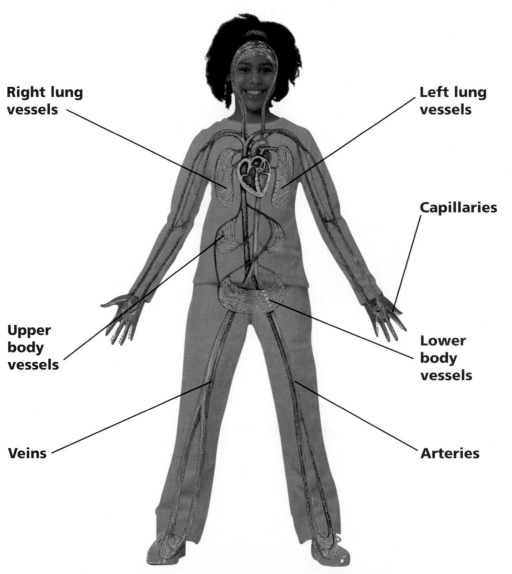

Right lung vessels

Left lung vessels

Capillaries

Upper body vessels

Lower body vessels

Veins

Arteries

The Heart

The heart has two sides, right and left, separated by a thick muscular wall. Each side has two chambers for blood. The upper chamber is the atrium. The lower chamber is the ventricle. Blood enters the heart through the vena cava. It leaves the heart through the aorta.

The pulmonary artery carries blood from the body into the lungs. Here carbon dioxide leaves the blood to be exhaled by the lungs. Fresh oxygen enters the blood to be carried to every cell in the body. Blood returns from the lungs to the heart through the pulmonary veins.

CARE!

- **Don't smoke. The nicotine in tobacco makes the heart beat faster and work harder to pump blood.**

- **Never take illegal drugs, such as cocaine or heroin. They can damage the heart and cause heart failure.**

How the Heart Works

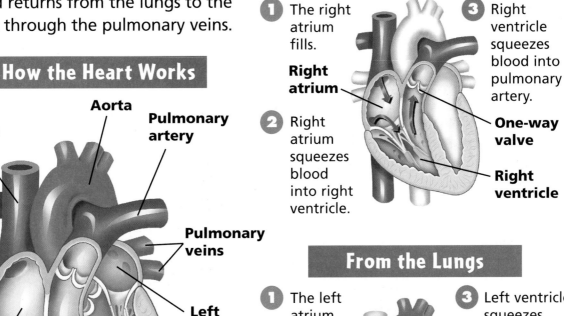

Vena cava

Aorta

Pulmonary artery

Pulmonary veins

Left atrium

Right atrium

Left ventricle

Right ventricle

Muscle wall

To the Lungs

1. The right atrium fills.

Right atrium

2. Right atrium squeezes blood into right ventricle.

3. Right ventricle squeezes blood into pulmonary artery.

One-way valve

Right ventricle

From the Lungs

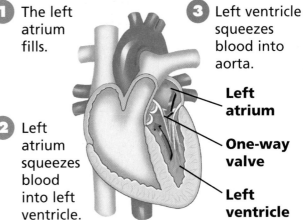

1. The left atrium fills.

2. Left atrium squeezes blood into left ventricle.

3. Left ventricle squeezes blood into aorta.

Left atrium

One-way valve

Left ventricle

The Respiratory System

The process of getting and using oxygen in the body is called respiration. When a person inhales, air is pulled into the nose or mouth. The air travels down into the trachea. In the chest the trachea divides into two bronchial tubes. One bronchial tube enters each lung. Each bronchial tube branches into smaller tubes called bronchioles.

At the end of each bronchiole are tiny air sacs called alveoli. The alveoli exchange carbon dioxide for oxygen.

Oxygen comes from the air we breathe. The main muscle that controls breathing is a dome-shaped sheet of muscle called the diaphragm.

To inhale, the diaphragm contracts and pulls down. To exhale, the diaphragm relaxes and returns to its dome shape.

CARE!

- **Don't smoke. Smoking damages your respiratory system.**

- **Exercise to strengthen your breathing muscles.**

- **If you ever have trouble breathing, tell an adult at once.**

Air Flow

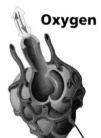

Carbon dioxide   **Oxygen**

Carbon dioxide diffuses into the alveoli. From there it is exhaled.

Capillary net

Alveoli

Fresh oxygen diffuses from the alveoli to the blood.

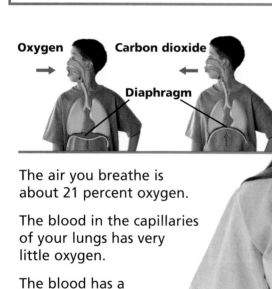

Oxygen →   **Carbon dioxide** ←

Diaphragm

Throat

Trachea

Lungs

The air you breathe is about 21 percent oxygen.

The blood in the capillaries of your lungs has very little oxygen.

The blood has a higher concentration of carbon dioxide than air.

Activity Pyramid

Physical fitness is the condition in which the body is healthy and works the best it can. The activity pyramid shows you the kinds of activities you should be doing to make your body more physically fit.

3–5 times a week Aerobic activities such as swimming, biking, climbing; sports activities such as basketball, handball

Occasionally
Inactive pastimes such as watching TV, playing board games, talking on the phone

2–3 times a week
Leisure activities such as gardening, golf, softball

Eating a variety of healthful foods and getting enough exercise and rest help people to stay healthy.

As people grow, the amounts and kinds of food and exercise the body needs may change.

Food Guide Pyramid

To make sure the body stays fit and healthy, a person needs to eat a balanced diet. The Food Guide Pyramid shows how many servings of each group a person should eat every day. Food provides energy and material for growth and repair of body parts. Vitamins and minerals keep the body healthy.

CARE!

- **Stay active every day.**
- **Eat a balanced diet.**
- **Drink plenty of water—6 to 8 large glasses a day.**

Fats, oils, and sweets
Use sparingly

Milk, yogurt, and cheese group
2–3 servings

Meat, dry beans, eggs, and nuts group
2–3 servings

Vegetable group
3–5 servings

Fruit group
2–4 servings

Bread, cereal, rice, and pasta group
6–11 servings

R 27

The Digestive System

Digestion is the process of breaking down food into simple substances the body can use. Digestion begins when a person chews food. Chewing breaks the food down into smaller pieces and moistens it with saliva.

Digested food is absorbed in the small intestine. The walls of the small intestine are lined with villi. Villi are tiny fingerlike projections that absorb digested food. From the villi the blood transports nutrients to every part of the body.

It is important to eat healthful foods. Avoid eating foods with caffeine, sugar, and fat as these foods often lack the nutrients the body needs. Stay away from alcohol and drugs as these substances damage the body.

CARE!

- Chew your food well.
- Drink plenty of water to help move food through your digestive system.

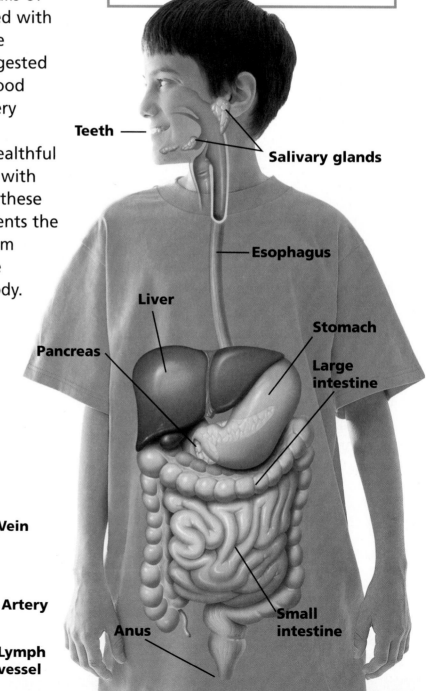

Teeth

Salivary glands

Esophagus

Liver

Pancreas

Stomach

Large intestine

Small intestine

Anus

Capillary

Villi

Vein

Artery

Lymph vessel

The Excretory System

Excretion is the process of removing waste products from the body. The liver filters wastes from the blood and converts them into urea. Urea is then carried to the kidneys for excretion.

The skin takes part in excretion when a person sweats. Glands in the inner layer of the skin produce sweat. Sweat is mostly water. Sweat tastes salty because it contains mineral salts the body doesn't need. There is also a tiny amount of urea in sweat.

Sweat is excreted onto the outer layer of the skin. Evaporation into the air takes place in part because of body heat. When sweat evaporates, a person feels cooler.

How You Sweat

Glands under your skin push sweat up to the surface, where it collects.

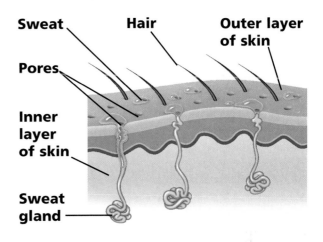

Sweat · Hair · Outer layer of skin · Pores · Inner layer of skin · Sweat gland

CARE!

- **Wash regularly to avoid body odor, clogged pores, and skin irritation.**

How Your Kidneys Work

1. Blood enters the kidney through an artery and flows into capillaries.

2. Sugars, salts, water, urea, and other wastes move from the capillaries to tiny nephrons.

3. Nutrients return to the blood and flow back out through veins.

4. Urea and other wastes become urine, which flows down the ureters.

5. Urine is stored in the bladder and excreted through the urethra.

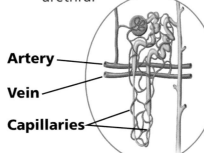

Artery · Vein · Capillaries

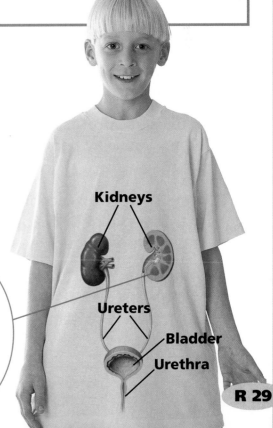

Kidneys · Ureters · Bladder · Urethra

The Nervous System

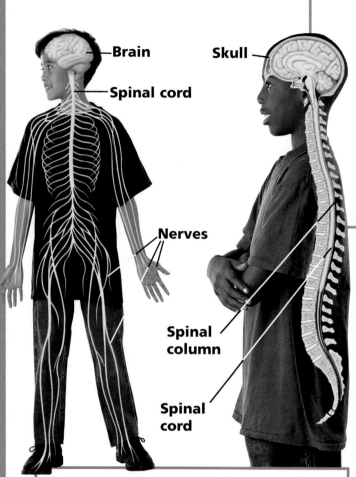

Brain
Spinal cord
Skull

Nerves

Spinal column

Spinal cord

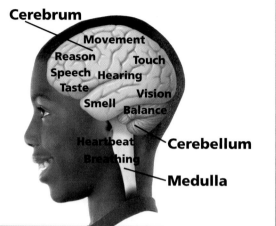

Cerebrum
Movement
Reason
Touch
Speech
Hearing
Taste
Smell
Vision
Balance
Heartbeat
Breathing
Cerebellum
Medulla

CARE!

- To protect the brain and spinal cord, wear protective headgear when you play sports or exercise.

- Stay away from alcohol, which is a depressant and slows down the nervous system.

- Stay away from drugs, such as stimulants, which can speed up the nervous system.

The nervous system has two parts. The brain and the spinal cord are the central nervous system. All other nerves are the outer nervous system.

The largest part of the brain is the cerebrum. A deep groove separates the right half, or hemisphere, of the cerebrum from the left half. Both sides of the cerebrum contain control centers for the senses.

The cerebellum lies below the cerebrum. It coordinates the skeletal muscles. It also helps in keeping balance.

The brain stem connects to the spinal cord. The lowest part of the brain stem is the medulla. It controls heartbeat, breathing, blood pressure, and the muscles in the digestive system.

The Endocrine System

Hormones are chemicals that control body functions. A gland that produces hormones is called an endocrine gland. Sweat from sweat glands flows out of tubes called ducts. Endocrine glands have no ducts.

The endocrine glands are scattered around the body. Each gland makes one or more hormones. Every hormone seeks out a target organ. This is the place in the body where the hormone acts.

Some Glands in the Endocrine System

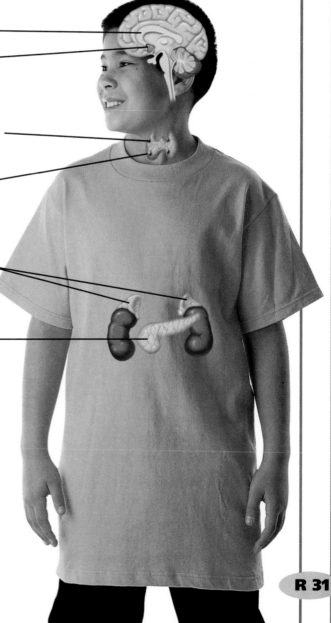

Hypothalamus

Pituitary gland

Parathyroid glands

Thyroid glands

Adrenal glands

Pancreas

CARE!

- Doctors can treat many diseases, such as diabetes, caused by endocrine glands that produce too little or too much of a hormone.

The Senses

Seeing

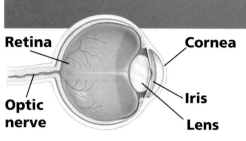

Retina

Cornea

Optic nerve

Iris

Lens

Light reflected from an object enters the eye and falls on the retina. Receptor cells change the light into electrical signals, or impulses. These impulses travel along the optic nerve to the vision center of the brain.

1 Light reflects off the tree and into your eyes.

2 The light passes through your cornea and the pupil in your iris.

3 Your eye bends the light so it hits your retina.

4 Receptor cells on your retina change the light into electrical signals.

5 The impulses travel along neurons in your optic nerve to the seeing center of your brain.

Hearing

1 Your outer ear collects sound waves.

2 They are funneled down your ear canal.

3 The eardrum vibrates.

Hammer

Eardrum

Anvil

Stirrup

4 Three tiny ear bones vibrate.

Auditory nerve

Cochlea

5 The cochlea vibrates.

6 Receptor cells inside your cochlea change.

Hearing center

7 The impulses travel along your auditory nerve to the brain's hearing center.

Sound waves enter the ear and cause the eardrum to vibrate. Receptor cells in the ear change the sound waves into impulses that travel along the auditory nerve to the hearing center of the brain.

CARE!

- Avoid loud music.
- Don't sit too close to the TV screen.

The Senses

Smelling

The sense of smell is really the ability to detect chemicals in the air. When a person breathes, chemicals dissolve in mucus in the upper part of the nose. When the chemicals come in contact with receptor cells, the cells send impulses along the olfactory nerve to the smelling center of the brain.

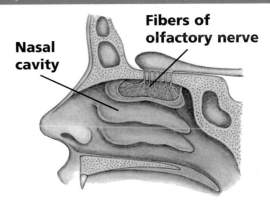

Nasal cavity

Fibers of olfactory nerve

Tasting

When a person eats, chemicals in food dissolve in saliva. Inside each taste bud are receptors that can sense the four main tastes—sweet, sour, salty, and bitter. The receptors send impulses along a nerve to the taste center of the brain. The brain identifies the taste of the food.

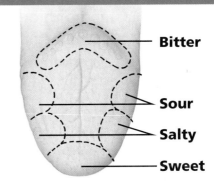

Bitter

Sour

Salty

Sweet

Touching

Receptor cells in the skin help a person tell hot from cold, wet from dry, and the light touch of a feather from the pressure of stepping on a stone. Each receptor cell sends impulses along sensory nerves to the spinal cord. The spinal cord then sends the impulses to the touch center of the brain.

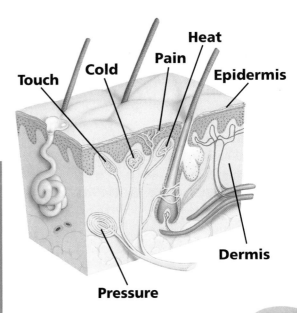

Touch   Cold   Pain   Heat   Epidermis

Dermis

Pressure

CARE!

- To prevent the spread of germs, always cover your mouth and nose when you cough or sneeze.

The Immune System

The immune system helps the body fight germs. Germs are tiny living things. The body is able to keep out harmful germs most of the time. Tears, saliva, and skin all help the body keep germs out. Sometimes germs get into the body. Usually white blood cells kill the germs before they can do any harm.

There are white blood cells in the blood vessels and in the lymph vessels. Lymph vessels are similar to blood vessels. Instead of blood, they carry lymph. Lymph nodes filter out harmful materials in the body. They also produce white blood cells to fight germs.

The white blood cells don't always get rid of the germs. Sometimes germs stay in the body and make it sick. When germs make the body sick, it is important to rest, eat healthful foods, and drink lots of water.

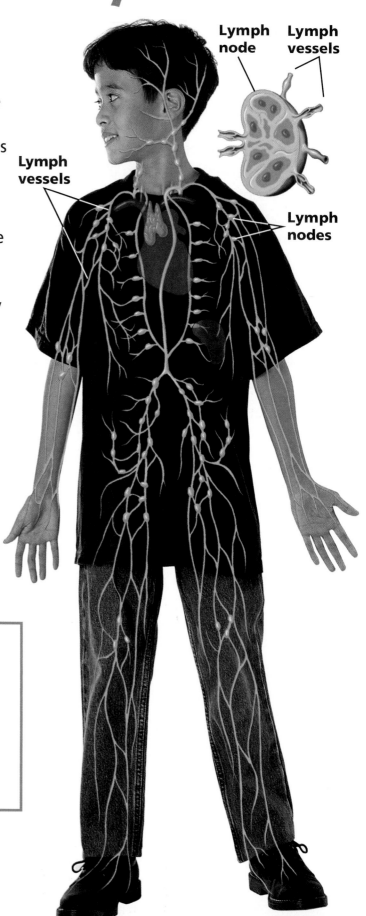

Lymph node

Lymph vessels

Lymph vessels

Lymph nodes

Lymph vessels run through your body to collect fluid and return it to the bloodstream.

CARE!

- Be sure to get immunized against common diseases.
- Keep cuts clean to prevent infection.

Nutrients

Nutrients are materials in foods that your body uses to grow and stay healthy. Without nutrients you could not grow, move, think, or even live. Most foods that provide you with nutrients come from plants and animals.

There are six kinds of nutrients and each helps your body in a different way. Some help you grow. Others help repair damaged tissues in your body. Some help your body function properly. Still others give you energy. The six kinds of nutrients found in foods are carbohydrates, vitamins, minerals, proteins, fats, and water. A balance of different foods will give your body the nutrients that it needs.

Carbohydrates

Your body needs a constant supply of energy to keep working. **Carbohydrates** are a main source of energy for your body. There are two kinds of carbohydrates. One is *starch* and the other is *sugar*.

Starches provide long-lasting energy. Your body is able to store energy from starches longer than many other nutrients. Foods with starches are rice, potatoes, bread, cereal, and pasta.

It is easy to see what starch looks like.

1. Cut a raw potato into several pieces on a chopping board. Be careful with the knife.
2. Look for a whitish, milky liquid on the potato and on the knife. That liquid contains starch.
3. Check other foods for starch, such as cooked pasta or cooked rice.

Energy from sugars doesn't last as long as energy from starches. Fruits such as apples and oranges are made of sugars.

How can you test a food to tell if it is made of sugars? One way is to test them for a chemical reaction.

Finding Sugar

1. Label each of five small paper cups *apple juice*, *orange juice*, *olive oil*, *milk*, and *water*. Pour a small amount of each liquid into its labeled cup.
2. Your teacher will give you one glucose strip for each cup. Do not touch the strips with your fingers. Use a tweezer to hold a glucose strip in each liquid for two seconds.
3. What happened to each of the test strips? Record your observations in a chart.

If the food you tested contains sugar, the yellow test strip will turn green. Which foods tested contained sugar?

Vitamins and Minerals

Vitamins keep your body tissues healthy and protect you from illness. They are found in foods that come from plants and animals.

Very small amounts of vitamins are present in many foods. But this is all our bodies need to grow and stay healthy. These are some common vitamins you can find in foods.

Minerals come from Earth. They are found in small amounts in foods that come from plants and animals. Minerals help your blood, muscles, and nervous system. They help your bones to grow and function.

Calcium is a mineral that builds strong teeth and bones. It's found in foods such as yogurt, milk, cheese, and green vegetables. *Iron* helps red blood cells. It can be found in meat, beans, fish, and whole grains. Your body uses *zinc* to grow and to heal wounds. It is found in meat, fish, and eggs.

Vitamin	Sources	Benefits
A	Milk, fruit, carrots, green vegetables	Keeps eyes, teeth, gums, skin and hair healthy
C	Citrus fruits, strawberries, tomatoes	Helps heart, cells, and muscle function
D	Milk, fish, eggs	Helps keep teeth and bones strong

Protein and Water

Two of the most important nutrients for any living thing are **protein** and water. Proteins are part of every living cell. They are needed by all organisms. They help your body grow and help repair body cells. Foods such as milk, dairy products, meats, fish, and nuts are good sources of protein.

You can see the protein in some foods. Have you ever cooked an egg in a frying pan? An egg is rich in protein. Even the colorless part of the egg is made of protein. As the egg cooks, you can see the protein in this part of the egg become white.

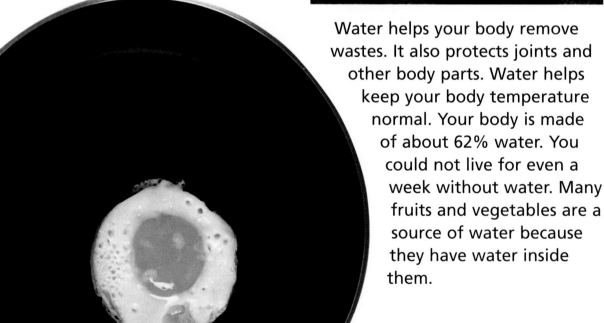

Water helps your body remove wastes. It also protects joints and other body parts. Water helps keep your body temperature normal. Your body is made of about 62% water. You could not live for even a week without water. Many fruits and vegetables are a source of water because they have water inside them.

Fats and Oils

Fats help your body to use other nutrients and to store vitamins. Fats keep your body warm and help brain cells and other body tissues work. Fats are found in meats, eggs, milk, butter, and nuts. Oils, such as those used in cooking, also contain fat.

While fat is needed for your body to work properly, it is needed only in small amounts. Some foods contain more fat than others.

Fat Check

Here's a way to find out which foods contain a lot of fats.

1. Gather various foods, such as a potato chip, cookie, carrot, and apple.
2. Cut a brown paper bag into 3-inch (7-cm) squares. Label each square with the name of one of the foods.
3. Rub some of each food on the square. Let the square dry.
4. Hold each square up to the light. What effect do the different foods have on the paper?

Foods that leave a greasy mark or stain on the paper contain a lot of fat, or oils. Too much fat in our bodies can cause health problems.

Calories

The chemical energy in foods is measured in a unit called a **calorie**. You take calories of food energy into your body by eating. Then you use, or burn, calories with everything you do. Breathing, eating, digesting food, walking, even doing your homework uses calories.

You can compare the chemical energy from different foods. Most food packages contain a label that gives information about the food's nutrients.

Nutrition Facts

Serving Size 9 Crackers (31g)
Servings Per Container About 13

Amount Per Serving

Calories 120   Calories from Fat 15

	% Daily Value*
Total Fat 1.5g	2 %
Saturated Fat .5g	3 %
Cholesterol 0mg	0 %
Sodium 210mg	9 %
Total Carbohydrate 25g	8 %
Dietary Fiber 1g	4 %
Sugars 0g	
Protein 2g	

Vitamin A 0%	•	Vitamin C 0%
Calcium 0%	•	Iron 6%

* Percent Daily Values are based on a 2,000 calorie diet. Your daily values may be higher or lower depending on your calorie needs:

	Calories:	2,000	2,500
Total Fat	Less than	65g	80g
Sat Fat	Less than	20g	25g
Cholesterol	Less than	300mg	300mg
Sodium	Less than	2,400mg	2,400mg
Total Carbohydrate		300g	375g
Dietary Fiber		25g	30g

Nutrition Facts

Serving Size 7 Crackers (31g)
Servings Per Container About 9

Amount Per Serving

Calories 140   Calories from Fat 45

	% Daily Value*
Total Fat 5g	8%
Saturated Fat 1g	5%
Polyunsaturated Fat 0g	
Monounsaturated Fat 1.5g	
Cholesterol 0mg	0%
Sodium 200mg	8%
Total Carbohydrate 21g	7%
Dietary Fiber 4g	14%
Sugars 0g	
Protein 3g	

Vitamin A 0%	•	Vitamin C 0%
Calcium 0%	•	Iron 8%

* Percent Daily Values are based on a 2,000 calorie diet. Your daily values may be higher or lower depending on your calorie needs:

	Calories:	2,000	2,500
Total Fat	Less than	65g	80g
Sat Fat	Less than	20g	25g
Cholesterol	Less than	300mg	300mg
Sodium	Less than	2,400mg	2,400mg
Total Carbohydrate		300g	375g
Dietary Fiber		25g	30g

Calorie Comparison

Compare the calories on the food labels of these two boxes of crackers.

1. Look at the serving size on both labels. How many grams (g) of crackers are in a serving of each? Are they equal?
2. Compare the calories listed on the labels.
3. Which cracker provides more chemical energy per serving? Which cracker contains more sugar?
4. Now compare labels on two different types of foods, such as cheese and butter. Be sure the serving sizes are equal before comparing calories.

FOLDABLES™

by Dinah Zike

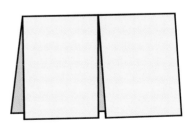

Folding Instructions

So how do you make a Foldables data organizer? The following pages offer step-by-step instructions—where and when to fold, where to cut—for making 11 basic Foldables data organizers. The instructions begin with the basic shapes, such as the hot dog fold, that were introduced on page xv.

Half-Book

Fold a sheet of paper ($8\frac{1}{2}$" x 11") in half.

1. This book can be folded vertically like a hot dog or …

2. … it can be folded horizontally like a hamburger.

Folded Book

1. Make a Half-Book.

2. Fold in half again like a hamburger.

This makes a ready-made cover and two small pages inside for recording information.

Two-Tab Book

Take a Folded Book and cut up the valley of the inside fold toward the mountain top.

This cut forms two large tabs that can be used front and back for writing and illustrations.

Pocket Book

1. Fold a sheet of paper ($8\frac{1}{2}$" x 11") in half like a hamburger.

2. Open the folded paper and fold one of the long sides up two inches to form a pocket. Refold along the hamburger fold so that the newly formed pockets are on the inside.

3. Glue the outer edges of the two-inch fold with a small amount of glue.

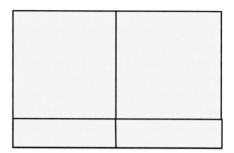

Shutter Fold

1. Begin as if you were going to make a hamburger, but instead of creasing the paper, pinch it to show the midpoint.

2. Fold the outer edges of the paper to meet at the pinch, or midpoint, forming a Shutter Fold.

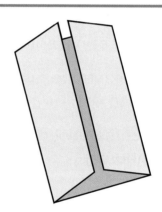

Trifold Book

1. Fold a sheet of paper ($8\frac{1}{2}$" x 11") into thirds.

2. Use this book as is, or cut into shapes.

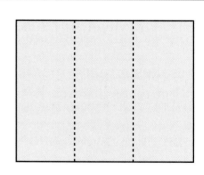

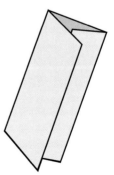

Three-Tab Book

1. Fold a sheet of paper like a hot dog.

2. With the paper horizontal and the fold of the hot dog up, fold the right side toward the center, trying to cover one half of the paper.

3. Fold the left side over the right side to make a book with three folds.

4. Open the folded book. Place one hand between the two thicknesses of paper and cut up the two valleys on one side only. This will create three tabs.

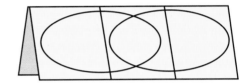

Layered-Look Book

1. Stack two sheets of paper ($8\frac{1}{2}$" x 11") so that the back sheet is one inch higher than the front sheet.

2. Bring the bottoms of both sheets upward and align the edges so that all of the layers or tabs are the same distance apart.

3. When all the tabs are an equal distance apart, fold the papers and crease well.

4. Open the papers and glue them together along the valley, or inner center fold, or staple them along the mountain.

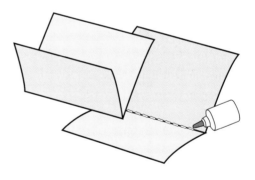

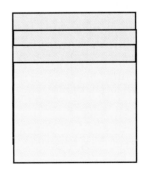

Four-Tab Book

1. Fold a sheet of paper ($8\frac{1}{2}$" x 11") in half like a hot dog.

2. Fold this long rectangle in half like a hamburger.

3. Fold both ends back to touch the mountain top or fold it like an accordion.

4. On the side with two valleys and one mountain top, make vertical cuts through one thickness of paper, forming four tabs.

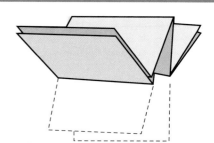
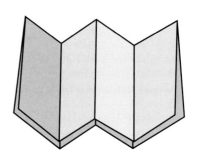

Four-Door Book

1. Make a Shutter Fold using 11" x 17" or 12" x 18" paper.

2. Fold the Shutter Fold in half like a hamburger. Crease well.

3. Open the project and cut along the two inside valley folds.

These cuts will form four doors on the inside of the project.

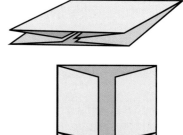

Folded Table or Chart

1. Fold the number of vertical columns needed to make the table or chart.

2. Fold the horizontal rows needed to make the table or chart.

3. Label the rows and columns.

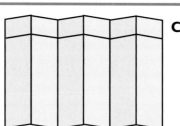

Chart

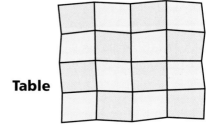

Table

Glossary

This Glossary will help you to pronounce and understand the meanings of the Science Words introduced in this book. The page number at the end of the definition tells where the word appears.

A

adaptation (ad′əp tā′shən) A special characteristic that helps an organism survive. (p. B50)

air (âr) A mixture of gases and dust. (p. D6)

air pressure (âr presh′ər) The force of air pushing down on Earth. (p. D9)

algae (al′jē) *pl. n., sing.* (-gə) Tiny one-celled organisms. (pp. B9, B27)

amber (am′bər) Hardened tree sap, often a source of insect fossils. (p. C22)

amphibian (am fib′ē ən) An animal that spends part of its life in water and part of its life on land. (p. A72)

anemometer (an′ə mom′i tər) A device that measures wind speed. (p. D25)

aqueduct (ak′wə dukt′) A pipe or channel for carrying water over long distances. (p. C32)

atmosphere (at′məs fîr′) Gases that surround Earth. (p. D6)

atom (at′əm) The smallest particle of matter. (p. F28)

axis (ak′sis) A real or imaginary line through the center of a spinning object. (p. D37)

PRONUNCIATION KEY

The following symbols are used throughout the Macmillan McGraw-Hill Science Glossaries.

a	at	e	end	o	hot	u	up	hw	**white**	ə	about
ā	ape	ē	me	ō	old	ū	use	ng	**song**		taken
ä	far	i	it	ôr	fork	ü	rule	th	**thin**		pencil
âr	care	ī	ice	oi	oil	u̇	pull	th	**this**		lemon
ô	law	îr	pierce	ou	out	ûr	turn	zh	measure		circus

′ = primary accent; shows which syllable takes the main stress, such as **kil** in **kilogram** (kil′ə gram′).

′ = secondary accent; shows which syllables take lighter stresses, such as **gram** in **kilogram.**

B

bacteria, (bak tîr′ē ə) One-celled living things. (p. B18)

barometer (bə rom′i tər) A device for measuring air pressure. (p. D24)

bird (bûrd) An animal that has a beak, feathers, two wings, and two legs. (p. A73)

bulb (bulb) The underground stem of such plants as onions and irises. (p. A30)

C

camouflage (kam′ə fläzh) An adaptation that allows animals to blend into their surroundings. (p. B52)

carbon dioxide and oxygen cycles (kär′bən dī ok′sīd and ok′sə jən sī′kəlz) The process of passing oxygen and carbon dioxide from one population to another in both water and land habitats. (p. B27)

carnivore (kär′nə vôr′) An animal that eats only other animals. (p. B20)

cast (kast) A fossil formed or shaped inside a mold. (p. C23)

cell (sel) **1.** The basic building block of life. (p. A10) **2.** A source of electricity. (p. F72)

cell membrane (sel mem′brān′) The thin outer covering of a cell. (p. A10)

cell wall (sel wôl) A stiff layer outside the cell membrane of plant cells. (p. A11)

chemical change (kem′i kəl chānj) A change that forms a different kind of matter. (p. F30)

chloroplast (klôr′ə plast′) One of the small green bodies inside a plant cell that makes foods for the plant. (p. A11)

circuit (sûr′kit) The path electricity flows through. (p. F72)

classify (klas′ə fī) To place materials that share properties together in groups. (pp. S3, A66)

communicate (kə mü′ni kāt′) To share information. (pp. S11, F20)

community (kə mü′ni tē) All the living things in an ecosystem. (p. B6)

competition (kom′pi tish′ən) The struggle among organisms for water, food, or other needs. (p. B42)

compound (kom′pound) Two or more elements put together. (p. F30)

compound machine (kom′pound mə shēn′) Two or more simple machines put together. (p. E57)

condense (kən dens′) v. To change from a gas to a liquid. (pp. C31, F17) —**condensation** (kon′den sā′shən) n. (p. D17)

conductor (kən duk′tər) A material that heat travels through easily. (p. F46)

conifer (kon′ə fər) A tree that produces seeds inside of cones. (p. A28)

conserve (kən sûrv′) To save, protect, or use something wisely without wasting it. (p. C34)

consumer (kən sü′mər) An organism that eats producers or other consumers. (pp. A40, B17)

crater (krā′tər) A hollow area in the ground. (p. D49)

cutting (kut′ing) A plant part from which a new plant can grow. (p. A30)

cytoplasm (sī′tə pla′zəm) A clear, jellylike material that fills both plant and animal cells. (p. A10)

D

decibel (dB) (des′ə bel′) A unit that measures loudness. (p. F69)

decomposer (dē′kəm pō′zər) An organism that breaks down dead plant and animal material. *Decomposers* recycle chemicals so they can be used again. (p. B18)

define based on observations (di fīn′ bāst ôn ob′zər vā′shənz) To put together a description that relies on examination and experience. (p. S5)

degree (di grē′) The unit of measurement for temperature. (p. F43)

PRONUNCIATION KEY

a at; ā ape; ä far; âr care; ô law; e end; ē me; i it; ī ice; îr pierce; o hot; ō old; ôr fork; oi oil; ou out; u up; ū use; ü rule; u̇ pull; ûr turn; hw white; ng song; th thin; th̲ this; zh measure; ə about, taken, pencil, lemon, circus

desert (dez′ərt) A hot, dry place with very little rain. (p. B55)

development (di vel′əp mənt) The way a living thing changes during its life. (p. A6)

distance (dis′təns) The length between two places. (p. E7)

E

earthquake (ûrth′kwāk) A sudden movement in the rocks that make up Earth's crust. (p. C72)

ecosystem (ek′ō sis′təm) All the living and nonliving things in an environment and all their interactions. (p. B6)

electric current (i lek′trik kûr′ənt) Electricity that flows through a circuit. (p. F72)

element (el′ə mənt) A building block of matter. (p. F28)

embryo (em′brē ō) A young organism that is just beginning to grow. (p. A26)

endangered (en dān′jərd) Close to becoming extinct; having very few of its kind left. (p. B64)

energy (en′ər jē) The ability to do work. (pp. A18, E39)

energy pyramid (en′ər jē pir′ə mid′) A diagram that shows how energy is used in an ecosystem. (p. B22)

environment (en vī′rən mənt) The things that make up an area, such as land, water, and air. (p. A8)

erosion (i rō′zhən) The carrying away of weathered materials. (p. C62)

evaporate (i vap′ə rāt′) *v.* To change from a liquid to a gas. (pp. C31, F17) —**evaporation** (i vap′ə rā′shən′) *n.* (p. D17)

experiment (ek sper′ə ment′) To perform a test to support or disprove a hypothesis. (pp. S7, A12)

extinct (ek stingkt) Died out, leaving no more of that type of organism alive. (p. B66)

F

fertilizer (fûr′tə līz) A substance added to the soil that is used to make plants grow. (p. C42)

first quarter (fûrst kwôr′tər) A phase of the Moon in which the right half is visible and growing larger. (p. D47)

fish (fish) An animal that lives its whole life in water. (p. A71)

flood (flud) A great rush of water over usually dry land. (p. C71)

flowering plant (flou'ər ing plant) A plant that produces seeds inside of flowers. (p. A28)

fog (fôg) A cloud that forms near the ground. (p. D14)

food chain (füd chān) A series of organisms that depend on one another for food. (p. B17)

food web (füd web) Several food chains that are connected. (p. B20)

force (fôrs) A push or pull, such as the one that moves a lever. (pp. E14, E44)

form a hypothesis (fôrm ə hī poth'ə sis) To make a statement that can be tested in answer to a question. (pp. S5, C64)

fossil (fos'əl) The imprint or remains of something that lived long ago. (p. C22)

freeze (frēz) To turn from water to ice. (p. F17)

friction (frik'shən) A force that occurs when one object rubs against another. (p. E26)

fuel (fū'əl) A substance burned for its energy. (p. C26)

fulcrum (fùl'krəm) The point where a lever turns or pivots. (p. E44)

full Moon (fùl mün) or **second quarter** (sek'ənd kwôr'tər) The phase of the Moon in which all of its sunlit half is visible from Earth. (p. D47)

fungi, (fun'jī) *pl. n., sing.* **fungus** (fung'gəs) One- or many-celled organisms that absorb food from dead organisms. (p. B18)

gas (gas) Matter that has no definite shape or volume. (p. F14)

PRONUNCIATION KEY

a at; ā ape; ä far; âr care; ô law; e end; ē me; i it; ī ice; îr pierce; o hot; ō old; ôr fork; oi oil; ou out; u up; ū use; ü rule; ù pull; ûr turn; hw white; ng song; th thin; <u>th</u> this; zh measure; ə about, taken, pencil, lemon, circus

germinate (jûr′mə nāt) To begin to grow, as when the right conditions allow a seed to develop. (p. A26)

glacier (glā′shər) A large mass of ice in motion. (p. C62)

gram (gram) A metric unit used to measure mass; 1,000 *grams* equals 1 kilogram. (p. F9)

gravity (grav′i tē) A pulling force between two objects, such as Earth and you. (p. E16)

groundwater (ground wô′tər) Water stored in the cracks of underground rocks and soil. (p. C33)

H

habitat (hab′i tat) The home of a living thing. (p. B7)

heat (hēt) A form of energy that makes things warmer. (p. F42)

herbivore (hûr′bə vôr′) An animal that eats only plants. (p. B20)

heredity (hə red′i tē) The passing of traits from parents to offspring. (p. A27)

hibernate (hī′bər nāt′) To rest or sleep through the cold winter. (p. A46)

host (hōst) An organism that a parasite lives with. (p. B31)

humus (hü′məs) Leftover decomposed plant and animal matter. (p. C14)

hurricane (hûr′i kān′) A violent storm with strong winds and heavy rains. (p. C70)

#

igneous rock (ig′nē əs rok) A "fire-made" rock formed from melted rock material. (p. C8)

imprint (im′print′) A shallow mark or print in a rock. (p. C23)

inclined plane (in klīnd′ plān) A flat surface that is raised at one end. (p. E54)

infer (in fûr′) To form an idea from facts or observations. (pp. S5, D20)

inherited trait (in her′i təd trāt) A characteristic that comes from parents. (p. A56)

inner planet (in'ər plan'it) Any of the four planets in the solar system that are closest to the Sun: Mercury, Venus, Earth, and Mars. (p. D56)

insulator (in'sə lā'tər) A material that heat doesn't travel through easily. (p. F46)

interpret data (in tûr'prit dā'tə) To use the information that has been gathered to answer questions or solve a problem. (p. S9)

K

key (kē) A table that shows what different symbols on a map stand for. (p. E10)

kilogram (kil'ə gram') A metric unit used to measure mass; 1 *kilogram* equals 1,000 grams. (p. F9)

L

landform (land'fôrm') A feature on Earth's surface. (p. C54)

last quarter (last kwôr'tər) or **third quarter** (thûrd kwôr'tər) The phase of the waning Moon in which the left half is visible but growing smaller. (p. D47)

leaf (lēf) A plant part that grows from the stem and helps the plant get air and make food. (p. A18)

learned trait (lûrnd trāt) Something that you are taught or learn from experience. (p. A56)

lens (lenz) A curved piece of glass. (p. D58)

lever (lev'ər) A straight bar that moves on a fixed point. (p. E44)

life cycle (līf sī'kəl) All the stages in an organism's life. (p. A26)

liquid (lik'wid) Matter that has a definite volume but not a definite shape. (p. F14)

liter (lē'tər) A metric unit used to measure volume. (p. F9)

load (lōd) The object that a lever lifts or moves. (p. E44)

PRONUNCIATION KEY

a at; ā ape; ä far; âr care; ô law; e end; ē me; i it; ī ice; îr pierce; o hot; ō old; ôr fork; oi oil; ou out; u up; ū use; ü rule; u̇ pull; ûr turn; hw white; ng song; th thin; th this; zh measure; ə about, taken, pencil, lemon, circus

loam (lōm) A kind of soil that contains clay, sand, silt, and humus. Plants grow well in loam. (p. C15)

luster (lus′tər) How an object reflects light. (p. F6)

M

machine (mə shēn′) A tool that makes work easier to do. (p. E44)

magnetism (mag′ni tiz′əm) The property of an object that makes it attract iron. (p. F26)

make a model (māk ə mod′əl) To make something to represent an object or event. (p. S7)

mammal (mam′əl) An animal with fur that feeds its young with milk. (p. A74)

map (map) A flat drawing that shows the positions of things. (p. E10)

mass (mas) The amount of matter in an object. (p. F7)

matter (mat′ər) Anything that takes up space and has mass. (pp. D16, F6)

measure (mezh′ər) To find the size, volume, area, mass, weight, or temperature of an object, or how long an event occurs. (pp. S9, C16)

melt (melt) To change from a solid to a liquid. (p. F17)

metal (met′əl) A shiny material found in the ground. (p. F26)

metamorphic rock (met′ə môr′fik rok) A rock that has changed form through squeezing and heating. (p. C9)

metamorphosis (met′ə môr′fə sis) A change in the body form of an organism. (p. A52)

microscope (mī′krə skōp′) A device that uses glass lenses to allow people to see very small things. (p. A10)

migrate (mī′grāt) To move to another place. (p. A46)

mimicry (mim′i krē) The imitation by one animal of the traits of another. (p. B53)

mineral (min′ə rəl) A naturally occurring substance, neither plant nor animal. (pp. A16, C6)

mixture (miks'chər) Different types of matter mixed together. The properties of each kind of matter in the mixture do not change. (p. F18)

mold (mold) An empty space in a rock that once contained an object such as a dead organism. (p. C23)

motion (mō'shən) A change in position. (p. E8)

mountain (moun'tən) The highest of Earth's landforms. *Mountains* often have steep sides and pointed tops. (p. C55)

N

natural resource (nach'ər əl rē'sôrs') A material on Earth that is necessary or useful to people. (p. C38)

nectar (nek'tər) The sugary liquid in flowers that lures insects that aid in pollination. (p. A28)

new Moon (nü mün) A phase of the Moon in which none of its sunlit half is visible from Earth. (p. D47)

newton (nü'tən) The unit used to measure pushes and pulls. (pp. E17, F10)

niche (nich) The job or role an organism has in an ecosystem. (p. B44)

nonrenewable resource (non'ri nü'ə bəl rē'sôrs') A resource that cannot be reused or replaced easily. (p. C41)

nucleus (nü'klē əs) The control center of a cell. (p. A11)

O

observe (əb sûrv') To use one or more of the senses to identify or learn about an object or event. (pp. S3, B56)

omnivore (om'nə vôr') An animal that eats both plants and animals. (p. B21)

PRONUNCIATION KEY

a at; ā ape; ä far; âr care; ô law; e end; ē me; i it; ī ice; îr pierce; o hot; ō old; ôr fork; oi oil; ou out; u up; ū use;
ü rule; u̇ pull; ûr turn; hw white; ng song; th thin; th this; zh measure; ə about, taken, pencil, lemon, circus

opaque (ō pāk′) A material that doesn't allow light to pass through. (p. F54)

orbit (ôr′bit) The path an object follows as it revolves around another object. (p. D38)

organ (ôr′gən) A group of tissues that work together. (p. A62)

organism (ôr′gə niz′əm) Any living thing. (p. A6)

outer planet (out′ər plan′it) Any of the five planets in the solar system that are farthest from the Sun: Jupiter, Saturn, Uranus, Neptune, and Pluto. (p. D56)

oxygen (ok′sə jən) A gas that is in air and water. (p. A19)

P

parasite (par′ə sīt′) An organism that lives in or on a host. (p. B31)

perish (per′ish) To fail to survive. (p. B63)

phase (fāz) An apparent change in the Moon's shape. (p. D46)

physical change (fiz′i kəl chānj) A change in the way matter looks that leaves the matter itself unchanged. (p. F16)

pitch (pich) How high or low a sound is. (p. F66)

plain (plān) Wide, flat lands. (p. C55)

planet (plan′it) Any of the nine large bodies that orbit the Sun. In order from the Sun outward, they are Mercury, Venus, Earth, Mars, Jupiter, Saturn, Uranus, Neptune, and Pluto. (p. D54)

pollen (pol′ən) A powdery material needed by the eggs of flowers to make seeds. (p. A28)

pollution (pə lü′shən) The adding of harmful substances to the water, air, or land. (p. C42)

population (pop′yə lā′shən) All the members of a single type of organism in an ecosystem. (p. B6)

position (pə zish′ən) The location of an object. (p. E6)

precipitation (pri sip′i tā′shən) Water in the atmosphere that falls to Earth as rain, snow, hail, or sleet. (p. D19)

predator (pred'ə tər) An animal that hunts other animals for food. (p. B28)

predict (pri dikt') To state possible results of an event or experiment. (pp. S7, D50)

prey (prā) The animals that predators eat. (p. B28)

prism (pri'zəm) A thick piece of glass that refracts light. (p. F57)

producer (prə dü'sər) An organism such as a plant that makes its own food. (p. B16)

property (prop'ər tē) Any characteristic of matter that you can observe. (p. F6)

pulley (pùl'ē) A simple machine that uses a wheel and a rope. (p. E48)

R

rain gauge (rān gāj) A device that measures how much precipitation has fallen. (p. D24)

ramp (ramp) Another name for an inclined plane. (p. E54)

recycle (rē sī'kəl) To treat something so it can be used again. (p. C44)

reduce (ri düs') To use less of something. (p. C44)

reflect (ri flekt') The bouncing of light off a surface. (p. F55)

refract (ri frakt') The bending of light as it passes through matter. (p. F56)

relocate (rē lō'kāt) To find a new home. (p. B63)

renewable resource (ri nü'ə bəl rē'sôrs') A resource that can be replaced or used over and over again. (p. C40)

reproduction (rē'prə duk'shən) The way organisms make more of their own kind. (p. A7)

reptile (rep'təl') An animal that lives on land and has waterproof skin. (p. A72)

PRONUNCIATION KEY

a at; ā ape; ä far; âr care; ô law; e end; ē me; i it; ī ice; îr pierce; o hot; ō old; ôr fork; oi oil; ou out; u up; ū use; ü rule; ù pull; ûr turn; hw white; ng song; th thin; <u>th</u> this; zh measure; ə about, taken, pencil, lemon, circus

reservoir (rez′ər vwär′) A storage area for fresh water supplies. (p. C32)

respond (ri spond′) To react to changes in the environment. (p. A8)

reuse (rē ūz′) To use something again. (p. C44)

revolve (ri volv′) To move around another object. (p. D38)

river (riv′ər) A large stream of water that flows across the land. (p. C55)

root (rüt) A plant part that takes in water and grows under the ground. (p. A17)

rotate (rō′tāt) To turn around. (p. D36)

S

sand dune (sand dün) A mound of windblown sand. (p. C55)

sapling (sap′ling) A very young tree. (p. A6)

satellite (sat′ə līt′) Any object that orbits another larger body in space. (p. D46)

scavenger (skav′ən jər) An animal that gets its food by eating dead organisms. (p. B29)

screw (skrü) An inclined plane wrapped into a spiral. (p. E56)

sedimentary rock (sed′ə men′tə rē rok) A kind of rock formed when sand, mud, or pebbles at the bottom of rivers, lakes, and oceans pile up. (p. C8)

seedling (sēd′ling) A young plant. (p. A27)

shelter (shel′tər) A place or object that protects an animal and keeps it safe. (p. A44)

simple machine (sim′pəl mə shēn′) A machine with few or no moving parts. (p. E44)

soil (soil) A mixture of tiny rock particles, minerals, and decayed plant and animal materials. (p. C14)

solar system (sō′lər sis′təm) The Sun and all the objects that orbit the Sun. (p. D54)

solid (sol′id) Matter that has a definite shape and volume. (p. F14)

solution (sə lū'shən) A kind of mixture in which one or more types of matter are mixed evenly in another type of matter. (p. F19)

speed (spēd) How fast an object moves over a certain distance. (p. E9)

sphere (sfîr) A body that has the shape of a ball or globe. (p. D36)

spore (spôr) One of the tiny reproductive bodies of ferns and mosses, similar to the seeds of other plants. (p. A30)

star (stär) A huge, hot sphere of gases, like the Sun, that gives off its own light. (p. D55)

stem (stem) A plant part that supports the plant. (p. A17)

switch (swich) A lever that opens or closes an electric circuit. (p. F73)

system (sis'təm) A group of parts that work together. (p. A62)

T

telescope (tel'ə skōp') A tool that gathers light to make faraway objects appear closer. (p. D58)

temperature (tem'pər ə cher) How hot or cold something is. (pp. D8, F43)

texture (teks'chər) How the surface of an object feels to the touch. (p. F6)

thermometer (thər mom'ə tər) An instrument used to measure temperature. (pp. D8, D24)

tissue (tish'ü) A group of cells that are alike. (p. A62)

tornado (tôr nā'dō) A violent, whirling wind that moves across the ground in a narrow path. (p. C70)

tuber (tü'bər) The underground stem of a plant such as the potato. (p. A30)

tundra (tun'drə) A cold, dry place. (p. B55)

PRONUNCIATION KEY

a at; ā ape; ä far; âr care; ô law; e end; ē me; i it; ī ice; îr pierce; o hot; ō old; ôr fork; oi oil; ou out; u up; ū use; ü rule; u̇ pull; ûr turn; hw white; ng song; th thin; <u>th</u> this; zh measure; ə about, taken, pencil, lemon, circus

U

use numbers (ūz num′bərz) To order, count, add, subtract, multiply, or divide to explain data. (p. S9)

use variables (ūz vâr′ē ə bəlz) To identify and separate things in an experiment that can be changed or controlled. (pp. S7, F58)

V

valley (val′ē) An area of low land lying between hills or mountains. (p. C55)

vibrate (vī′brāt) To move back and forth quickly. (p. F64)

volcano (vol kā′nō) An opening in the surface of Earth. (p. C73)

volume (vol′ūm) **1.** A measure of how much space matter takes up. (p. F7) **2.** How loud or soft a sound is. (p. F67)

W

water cycle (wô′tər sī′kəl) The movement of Earth's water over and over from a liquid to a gas and from a gas to a liquid. (pp. C31, D19)

water vapor (wô′tər vā′pər) Water in the form of a gas in Earth's atmosphere. (p. D17)

weather (weth′ər) The condition of the atmosphere at a given time and place. (p. D6)

weather vane (weth′ər vān) A device that indicates the direction of the wind. (p. D25)

weathering (weth′ər ing) The process that causes rocks to crumble, crack, and break. (p. C60)

wedge (wej) Two inclined planes placed back-to-back. (p. E55)

weight (wāt) The measure of the pull of gravity between an object and Earth. (p. E17)

wheel and axle (hwēl and ak′səl) A wheel that turns on a post. (p. E47)

wind (wind) Moving air. (p. D10)

windlass (wind′ləs) A wheel and axle machine that is turned by a hand crank to lift a bucket in a well. (p. E47)

work (wûrk) The force that changes the motion of an object. (p. E38)

Index

* Indicates an activity related to this topic.

R 59

* Indicates an activity related to this topic.

* Indicates an activity related to this topic.

* Indicates an activity related to this topic.

* Indicates an activity related to this topic.

Credits

Cover Photos: c. Leeson Photography; bkgd. Roderick Chen/ Superstock; spine Leeson Photography. Back Cover: bkgd. Roderick Chen/Superstock; t.l. Clive Druett/Papilio/CORBIS; t.r. Tim Flach/Stone/ Getty Images; c.l. Donovan Reese/Stone/Getty Images; c.r. Earth Satellite Corporation/Science Photo Library/Photo Researchers, Inc.; b.l. James Marshall/The Stock Market/CORBIS; b.r. SuperStock. Endpaper: Roderick Chen/Superstock.

Photography Credits: All photos are by Macmillan/McGraw-Hill (MMH) and Ken Karp for MMH, Ray Boudreau for MMH, Dan Howell for MMH, David Waitz for MMH, Ron Tanaka for MMH, Dave Mager for MMH, Richard Hutchings for MMH and John Serafin for MMH except as noted below:

i: bkgd. Roderick Chen/Superstock; t.l, b.l. Leeson Photography. iii: Leeson Photography. iv: t. NASA/CORBIS; c., b. Courtesy Sally Ride; bkgd. Taxi/Getty Images. v: l. Mervyn Rees/Alamy; r. James L. Amos/ CORBIS. vi: l. Victoria McCormick/Animals Animals; b. PhotoDisc/Getty Images. vii: i. inset Tim Davis/Stone/Getty Images; l. bkgd. Christer Fredriksson/Natural Selection Stock Photography; b. Runk/ Schoenberger/Grant Heilman Photography, Inc.; ants PhotoDisc/Getty Images. viii: l. Donovan Reese/Stone/Getty Images; b.c. Francois Gohier/Photo Researchers, Inc.; b.r. American Museum of Natural History. ix: l. Earth Satellite Corporation/Science Photo Library/Photo Researchers, Inc.; c. NASA/Photo Researchers, Inc.; r. John Sanford/ Photo Researchers, Inc. x: l. Ron Stroud/Masterfile; b. Peter Weimann/ Animals Animals/Earth Scenes. xi: l. Superstock; r. D. Boone/CORBIS. xiv: l. Roderick Chen/Superstock. xvi: b.r.: PhotoDisc/Getty Images. SO: Danny Lehman/CORBIS. S0-S1: James L. Amos/CORBIS. S2-S3: Tom Bean/CORBIS. S4: Mervyn Rees/Alamy. S5: DK Images. S6-S7: Ira Block/ National Geographic. S8-S9: Robert Campbell/CORBIS. S10-S11: Richard Cummins/CORBIS. AO: Clive Druett/Papilio/CORBIS. A0-A1: Victoria McCormick/Animals Animals. A2-A3: Alan Oddie/PhotoEdit. A4-A5: Douglas Peebles/CORBIS. A6: l. Tony Wharton/CORBIS; t., b. Terry Eggers/The Stock Market/CORBIS; r. The Stock Market/CORBIS. A6-A7: Frank Siteman/Stock Boston. A7: PhotoDisc/Getty Images. A8: t. Gerard Fuehrer/DRK Photo; b. SuperStock. A9: l. Norbert Wu/Peter Arnold, Inc.; c. Secret Sea Visions/Peter Arnold, Inc.; b. Joe McDonald/ Visuals Unlimited, Inc. A10: t. Kent Wood/Photo Researchers, Inc.; b. Dwight R. Kuhn. A11: Moredun Animal Health LTD/Science Photo Library/Photo Researchers, Inc. A13: Carl Roessler/Animals Animals/ Earth Scenes. A14: t.l. Doug Peebles/Panoramic Images; r. Mark Segal/Panoramic Images. A14-A15: bkgd. Ryan McVay/PhotoDisc/Getty Images; c. Allen Prier/Panoramic Images. A16: bkgd. Runk/ Schoenberger/Grant Heilman Photography; inset E. Webber/Visuals Unlimited. A17: b. Jim Zipp/Photo Researchers, Inc.; r. Jenny Hager/The Image Works. A18: l. Runk/Schoenberger/Grant Heilman Photography; r. C.G. Van Dyke/Visuals Unlimited. A19: t. Dave M. Phillips/Visuals Unlimited. A20: t. Bill Beatty/Visuals Unlimited; b. Pat O'Hara/DRK Photo. A22: l. Stan Osolinski/Dembinsky Photo Associates; r. Larry West/Taxi/Getty Images. A22-A23: Randy Green/Taxi/Getty Images. A23: t. John M. Roberts/The Stock Market/CORBIS; b. J. H. Robinson/ Photo Researchers, Inc. A24-A25: Neil Gilchrist/Panoramic Images. A26: t. D. Gavagnaro/Visuals Unlimited; t.c.r. Kevin Collins/Visuals Unlimited; c.r. Tony Freeman/PhotoEdit; b. Inga Spence/Tom Stack & Associates. A26-A27: Inga Spence/Visuals Unlimited. A27: D. Gavagnaro/Visuals Unlimited. A29: Gerald and Buff Corsi/Visuals Unlimited. A30: t., c. David Young-Wolff/PhotoEdit; b.l. Ed Reschke/ Peter Arnold, Inc.; b.r. Jeff J. Daly/Visuals Unlimited. A32: Ed Galindo. A32-A33: l., c., t. Ed Galindo; b.r. C Squared Studios/PhotoDisc/Getty Images. A36-A37: Stephen J. Krasemann/Photo Researchers, Inc. A38-A39: Jade Albert/FPG International/Getty Images. A40: t. Fritz Polikng/Bruce Coleman, Inc.; b.l. Joe McDonald/DRK Photo; b.r. Dale E. Boyer/Photo Researchers, Inc. A41: Kevin Schafer/Peter Arnold, Inc. A42: inset W. Gregory Brown/Animals Animals. A42-A43: Michael S. Nolan/Tom Stack & Associates. A44: t. Eric & David Hosking/CORBIS; c. John Cancalosi/DRK Photo; b. Ted Levine/Animals Animals. A45: t. Zoran Milich/Allsport USA/Getty Images; b. Mark Newman/Bruce Coleman, Inc. A 46: t. David Madison/Bruce Coleman, Inc.; c. Runk/ Schoenberger/Grant Heilman Photography; b. John Cancalosi/DRK Photo. A48: t. Robert P. Carr/Bruce Coleman, Inc.; b. Skip Moody/ Dembinsky Photo Associates. A48-A49: Ken Lucas/Visuals Unlimited. A49: t.l., t.r., b. Robert P. Carr/Bruce Coleman, Inc.; c. Jon Dicus. A50-A51: Tim Davis/Photo Researchers, Inc. A52: t. Dwight R. Kuhn; b. Arthur Morris/Visuals Unlimited. A53: t.l. Gelnn M. Oliver/Visuals Unlimited; t.c. Pat Lynch/Zipp/Photo Researchers, Inc.; t.r. Robert P. Carr/Bruce Coleman, Inc.; c.l. Nuridsany et Perennou/Zipp/Photo Researchers, Inc.; c.r. Robert L. Dunne/Bruce Coleman, Inc.; b.l. Sharon Cummings/Dembinsky Photo Associates; b.r. John Mielcarek/ Dembinsky Photo Associates. A54: t. SuperStock; c. Lynn Rogers/Peter Arnold, Inc.; b .l. Erwin and Peggy Bauer/Bruce Coleman, Inc.; b.r. Pat

and Tom Leeson/Photo Researchers, Inc. A55: t.l. Cabisco/Visuals Unlimited; t.c.l. E.A. Janes/Bruce Coleman, Inc.; t.r. Lindholm/Visuals Unlimited; b.c.l. Fred Breummer/DRK Photo; b.r. Dave B. Fleetham/ Visuals Unlimited; b.l. M H Sharp/Photo Researchers, Inc. A56: l. George Shelley/The Stock Market/CORBIS; r. Richard Hutchings/ PhotoEdit. A58: Bill Banaszewski/Visuals Unlimited. A58-A59: J.C. Carton/Bruce Coleman, Inc. A60-A61: Robert Maier/Animals Animals. A62: l. M.I. Walker/Science Source/Photo Researchers, Inc. A63: t. R. Dowling/Animals Animals/Earth Scenes; b. Joe McDonald/Animals Animals/Earth Scenes. A64: Robert Winslow. A64-A65: Tom Brakefield/ CORBIS. A65: t. James Watt/Animals Animals/Earth Scenes; b. Jeff Rotman/Jeff Rotman Photography. A66: PhotoDisc/Getty Images. A68-A69: VCG/FPG International/Getty Images. A69: t.r. Stephen Dalton/Animals Animals/Earth Scenes; t.c.r. Tony Wharton/CORBIS; b.c.r. Brian Parker/Tom Stack & Associates; b.r. Lisa and Mike Husar/ DRK Photo; c.l. EyeWire; b.l. Kichen and Hurst/Tom Stack & Associates; c. G.W.Willis/Animals Animals/Earth Scenes. A70: l. Darryl Torckler/ Stone/Getty Images; r. Rob Simpson/Visuals Unlimited. A71: t. Breck P. Kent/Animals Animals/Earth Scenes; inset George Bernard/Animals Animals/Earth Scenes. A72: t. Jane Burton/Bruce Coleman, Inc.; b. E.R. Degginger/Animals Animals/Earth Scenes. A73: l. S. Nielson/ DRK Photo; r. Robert Winslow. A74: Jeff Rotman/Jeff Rotman Photography. A74-A75: Dave Watts/Tom Stack & Associates. A75: t. Erwin & Peggy Bauer/Bruce Coleman, Inc.; b. Lynn M. Stone/Bruce Coleman Inc. A77: SuperStock. A78: Photo courtesy of Dan Lausser/ University of Wisconsin-Madison. A78-A79: c. Photo courtesy Jeff Miller/University of Wisconsin-Madison; bkgd. StockTrek/Photodisc Green/Getty Images, Inc. A80: l. Jeff J. Daly/Visuals Unlimited; r. David Young-Wolff/PhotoEdit. BO: Tim Flach/Stone/Getty Images. B0-B1: Christer Fredriksson/Natural Selection Stock Photography. B1: Tim Davis/Stone/Getty Images. B2-B3: Raymond Gehman/CORBIS. B4-B5: Johnny Johnson/Animals Animals/Earth Scenes. B5: PhotoDisc/Getty Images. B6: Nicholas DeVore/Stone/Getty Images. B6-B7: Joseph Van Os/The Image Bank/Getty Images. B12: inset Lance Nelson/The Stock Market/CORBIS; c. Jeff Greenberg/Visuals Unlimited; b. Jeff Greenberg/ PhotoEdit. B13: t. Gerard Lacz/Peter Arnold, Inc.; b. PhotoDisc/Getty Images. B14-B15: L. Lenz/Natural Selection. B16: t. Kim Taylor/Dorling Kindersley Ltd. B18: t. SuperStock; b.l. Michael P. Gadomski/Photo Researchers, Inc. B18-B19: Dwight Kuhn Photography. B22: t. John Warden/Stone/Getty Images; t.c. Tom J. Ulrich/Visuals Unlimited; b.c. John Shaw/Bruce Coleman, Inc.; b. Runk/Schoenberger/Grant Heilman Photography, Inc. B24-B25: Michael Simpson/FPG International/Getty Images. B26: William H. Mullins/Photo Researchers, Inc. B26-B27: bkgd. Kent Foster/Photo Researchers, Inc. B28: t. Kim Taylor/Dorling Kindersley Ltd.; b. John Shaw/Bruce Coleman, Inc.; branch Kim Taylor/ Dorling Kindersley Ltd.; b. Arthur Morris/The Stock Market/CORBIS. B28-B29: inset Kim Taylor/Dorling Kindersley Ltd.; bkgd. M. C. Chamberlain/DRK Photo. B29: t. Jeremy Woodhouse/PhotoDisc/Getty Images; b. Jerry Young/Dorling Kindersley Ltd. B30: l. Nawrocki Stock Photo; b. Carl Roessler/Bruce Coleman, Inc. B30-B31: S. Dimmitt/Photo Researchers, Inc. B31: t. James H. Robinson/Photo Researchers, Inc.; b.l. Patricia Doyle/Stone/Getty Images; b.r. Runk/Schoenberger/Grant Heilman Photography, Inc. B32-B33: Kim Taylor/Dorling Kindersley Ltd. B34: l. Trevor Barrett/Animals Animals/Earth Scenes; b. George D. Lepp/Photo Researchers, Inc. B34-B35: John Elk III. B35: Kjell B. Sandved/Visuals Unlimited. B37: l., c. Runk/Schoenberger/Grant Heilman Photography, Inc.; r. Arthur Morris/Visuals Unlimited. B38-B39: Robert Winslow. B40-B41: The Stock Market/CORBIS. B42: t. Stephen Dalton/Animals Animals/Earth Scenes; b. Richard Day/ Panoramic Images. B43: t.l. Steve Maslowski/Visuals Unlimited; t.r. James P. Rowan/DRK Photo; b.l. George D. Dodge/Bruce Coleman, Inc.; b.r. Michael Dwyer/Stock Boston. B44: t. Gail Shumway/FPG International/Getty Images. B46: Paul McCormick/Getty Images. B47: Gil T Friedman. B48: Richard & Susan Day/Animals Animals/Earth Scenes. B48-B49: bkgd. Johnny Johnson/DRK Photo; c. Tom and Pat Leeson/DRK Photo. B50: t. Gail Shumway/FPG International/Getty Images; c. Jack Jeffrey/Photo Resource Hawaii; b. John Cancalosi/DRK Photo. B50-B51: t. Francis/Donna Caldwell/Visuals Unlimited; b. Kim Taylor/Bruce Coleman, Inc. B51: Heather Angel/Natural Visions. B52: t. Gregory Ochoki/Photo Researchers, Inc.; b. Breck P. Kent/Animals Earth Scenes. B53: t. Stephen J. Krasemann/DRK Photo; b.l. John Eastcott/Yva Momatiuk/DRK Photo; b.r. A. Cosmos Blank/Photo Researchers, Inc. B54: t. Zig Leszczynski/Animals Animals/Earth Scenes; b. Michael Fogden/DRK Photo. B54-B55: Michael Fogden/DRK Photo. B55: t.l. Pat O'Hara/DRK Photo; t.c. Don Enger/Animals Animals/Earth Scenes; t.r. Richard Kolar/Animals Animals/Earth Scenes; b.l. Jim Steinberg/Photo Researchers, Inc.; b.r. Stephen J. Krasemann/Photo Researchers, Inc. B57: Chris Johns/National Geographic . B58-B59: Gary Braasch/CORBIS. B60: t. Charles Palek/Earth Scenes; b. Pat and Tom Leeson/Photo Researchers, Inc. B60-B61: Brett Baunton. B61: Wayne

Hacker/Alamy. B62: l. Kent and Donna Dannen/Photo Researchers, Inc.; t.r. Diana L. Stratton/Tom Stack & Associates; t.c.r. Doug Sokell/Visuals Unlimited; b.c.r. Sharon Gerig/Tom Stack & Associates; b.r. Pat and Tom Leeson/DRK Photo. B63: t. Joe & Carol McDonald/Visuals Unlimited; b. Stephen J. Krasemann/DRK Photo. B64: t. M.C. Chamberlain/DRK Photo; b.l. Erwin and Peggy Bauer/Bruce Coleman Inc.; b.r. G. Prance/Visuals Unlimited. B65: M.C. Chamberlain/DRK Photo. B66: t. Stephen J. Krasemann/DRK Photo; b. Science VU/Visuals Unlimited. B69: l. Pat and Tom Leeson/Photo Researchers, Inc.; r. Robert Madden/National Geographic . B70: Smithsonian. B70-B71: b. PhotoDisc/Getty Images; bkgd. Cartesia/PhotoDisc. B71: t. Martin Harvey/CORBIS; b. Royalty-Free/CORBIS. CO: Donovan Reese/Stone/Getty Images. C0-C1: David Muench/Stone/Getty Images. C2-C3: David Muench/CORBIS. C4-C5: Chip Porter/Stone/Getty Images. C6: t. Joyce Photographics/Photo Researchers, Inc.; b.l. Bill Bachmann/Index Stock Imagery; b.r. Runk/Schoenberger/Grant Heilman Photography. C7: t., c. Tom Pantages. C8: Adam G. Sylvester/Photo Researchers, Inc. C9: t.l. Joyce Photographics/Photo Researchers, Inc.; t.r. Runk/Schoenberger/ Grant Heilman Photography; b.l. Charles R. Belinky/Photo Researchers, Inc. C10: t. Frederik D. Bodin/Stock Boston; c. Erich Lessing/Art Resource; b. Boleslaw Edelhajt/Gamma-Liaison/Getty Images. C12-C13: Bo Brannhage/Panoramic Images. C14-C15: Jeff Lepore/Panoramic Images. C15: l. Stephen Ogilvy for MMH. C17: Runk/Schoenberger/ Grant Heilman Photography. C18: t. The National Archives/CORBIS; b. G. Buttner/Okapia/Photo Researchers, Inc.; inset, c.r. Roy Morsch/The Stock Market/CORBIS. C19: bkgd. Arthur C. Smith/Grant Heilman Photography; t. Roy Morsch/The Stock Market/CORBIS. C20-C21: bkgd. Jeff J. Daly/Visuals Unlimited. C22: Tom Bean/Stone/Getty Images. C23: t. Runk/ Schoenberger/Grant Heilman Photography. C24: t. Louis Psihoyos/ MATRIX; inset Mehau Kulyk/Photo Researchers, Inc.; b. Francois Gohier/Photo Researchers, Inc. C24-C25: Biophoto Associates/Photo Researchers, Inc. C25: Stephen J. Krasemann/DRK Photo. C26: Ray Ellis/Photo Researchers, Inc. C28-C29: F. Stuart Westmorland/Photo Researchers, Inc. C30-C31: Tom Van Sant/Photo Researchers, Inc. C32: t. Davis Barber/PhotoEdit; b. C.C. Lockwood/DRK Photo. C36-C37: Grant Heilman/Grant Heilman Photography. C37: Emma Lee/Life File/PhotoDisc/Getty Images. C38-C39: t. David R. Frazier/Photo Researchers, Inc.; b. Charles Mauzy/Natural Selection. C39: John Elk III. C40: t. Don and Pat Valenti/DRK Photo; b. Gary Gray/DRK Photo. C41: t. American Museum of Natural History; inset Will and Deni McIntyre/ Photo Researchers, Inc.; b. George Gerster/Photo Researchers, Inc. C42: t. Ruth Dixon/Stock Boston; b. David Ulmer/Stock Boston. C42-C43: Simon Fraser/Science Photo Library/Photo Researchers, Inc. C44: t.l. Larry Lefever/Grant Heilman Photography; t.c.l. EyeWire; b.c.l. RJ Erwin/DRK Photo; b.l., b.r. Tony Freeman/PhotoEdit. C46: t. David Young-Wolff/PhotoEdit; c. Chromosohm/Sohm/Stock Boston. C47: t. Spencer Grant/PhotoEdit; b. Bonnie Kaman/PhotoEdit. C50-C51: Addison Geary/Stock Boston. C52-C53: Allen Prier/Panoramic Images. C53: t. Peter Miller/Panoramic Images; t.c. Richard Sisk/Panoramic Images; c. Mark Heifner/Panoramic Images; b.c. Kim Heacox/Stone/ Getty Images; b.r. Don Pitcher/Stock Boston; b.l. Jack Krawczyk/ Panoramic Images. C56: t.l. Jim Wiebe/Panoramic Images; t.r. Peter Pearson/Stone/Getty Images; b.l. Richard Sisk/Panoramic Images; b.c. Mark Heifner/Panoramic Images; b.r. Tom Bean/Stone/Getty Images. C57: t. Jack Krawczyk/Panoramic Images. C58-C59: David L. Brown/ Panoramic Images. C60: t. Michael P. Gadomski/Photo Researchers, Inc.; b. John Anderson/Animals Animals/Earth Scenes. C62: t. Thomas Fletcher/Stock Boston; inset PhotoDisc/Getty Images; b. Jeff Greenberg/PhotoEdit. C63: t. Kathy Ferguson/PhotoEdit; b. Runk/ Schoenberger/Grant Heilman Photography. C65: PhotoEdit. C66: Adam Jones/Photo Researchers, Inc. C66-C67: The National Archives/CORBIS. C67: t. W. E. Ruth/Bruce Coleman, Inc.; inset Pat Armstrong/Visuals Unlimited; c. Sylvan H. Wittaver/Visuals Unlimited; b. John Sohlden/ Visuals Unlimited. C68-C69: David Young-Wolff/PhotoEdit. C70-C71: t. Ana Laura Gonzalez/Animals Animals/Earth Scenes; b. Art Montes De Oca/FPG International/Getty Images. C72: t. David Bartruff/FPG International/Getty Images; b. Will & Deni McIntyre/Photo Researchers, Inc. C73: G. Brad Lewis/Stone/Getty Images. C74: t. David Weintraub/Stock Boston; b. Arthur Rothstein/Library of Congress/ Archive Photos/Getty Images. C78: M. Olsen/College of William and Mary. C78-C79: inset Woods Hole Oceanographic Institute; bkgd. Ralph White/CORBIS. DO: bkgd. Earth Satellite Corporation/Science Photo Library/Photo Researchers, Inc. D0: t. John Sanford/Science Photo Library/Photo Researchers, Inc. D0-D1: bkgd. Science Photo Library/Photo Researchers, Inc; inset Photo Library International/Photo Researchers, Inc. D2-D3: Jack Krawczyk/Panoramic Images. D4-D5: Ariel Skelley/The Stock Market/CORBIS. D8-D9: Didier Givois/Photo Researchers, Inc. D10: David Young-Wolff/PhotoEdit. D11: l. Barbara Stotzen/PhotoEdit; r. D. Boone/Corbis. D12: Jim Reed/CORBIS. D13: t. Jim Reed/Photo Researchers, Inc.; b. StockTrek/Photodisc Green/ Getty Images. D14-D15: Clifford Paine/CORBIS. D16: t. Myrleen Ferguson/PhotoEdit; c. Paul Silverman; b. Michael Newman/PhotoEdit. D17: t. Diane Hirsch/Fundamental Photographs; b. Jeff Greenberg/

Peter Arnold, Inc. D21: P. Quittemelle/Stock Boston. D22-D23: Bob Krist/CORBIS. D24: t., b.l. Tom Pantages; b.r. Jeff J. Daly/Stock Boston. D25: t. Charles D. Winters/Photo Researchers, Inc.; b. Tony Freeman/ PhotoEdit. D27: b. NOAA/Science Photo Library/Photo Researchers, Inc. D28: Michael P. Gadomski/Photo Researchers, Inc. D28-D29: F. Stuart Westmorland/Photo Researchers, Inc. D32-D33: Michael Hovell/Index Stock Imagery. D34-D35: Robert Mathena/Fundamental Photographs. D37: t. Ken Lucas/Visuals Unlimited; b. Thomas Barbudo/Panoramic Images. D41: Bob Daemmrich/Stock Boston. D42: Roger Ressmeyer/ CORBIS. D42-D43: Roger Ressmeyer/CORBIS. D43: John Sanford/Photo Network/Alamy. D44-D45: Peter Menzel/Stock Boston. D46: Frank Cara/Bruce Coleman, Inc. D47: John Sanford/Photo Researchers, Inc. D48: NASA/Science Source/Photo Researchers, Inc. D49: l. Mark E. Gibson/Visuals Unlimited; r. NASA/Science Photo Library/Photo Researchers, Inc. D52-D53: Frank Zullo/Photo Researchers, Inc. D56: t. U.S .Geological Survey/Photo Researchers, Inc.; c. NASA/Mark Marten/Photo Researchers, Inc.; b.l. Stock Boston; b.r. NASA/Tom Pantages. D56-D57: Ross Ressmeyer/NASA/CORBIS. D57: t.r. NASA/ Photo Researchers, Inc.; c.l. Space Telescope Space Institute/Photo Researchers, Inc.; c.r. NASA/Tom Pantages; b. Space Telescope Space Institute/Photo Researchers, Inc. D58: t. Tony Freeman/PhotoEdit. D62: t. National Weather Service; c., b. NOAA/AFP/Getty Images. D62-D63: bkgd. Don Farrall/PhotoDisc/Getty Images. D64: b. bkgd. Bruce Heinemann/PhotoDisc/Getty Images. EO: James Marshall/The Stock Market/CORBIS. E0-E1: Ron Stroud/Masterfile. E2-E3: Bernard Asset/Photo Researchers, Inc. E4-E5: S. Dalton/Photo Researchers, Inc. E5: t. Will Hart/PhotoEdit. E6: leaves Foodpix; snails Gregory K. Scott/Photo Researchers, Inc.; b.l., b.r. Zoran Milich/Allsport USA/Getty Images. E8: t. Robert Winslow; b. Fritz Polking/Peter Arnold, Inc. E8-E9: bkgd. The Stock Market/CORBIS. E9: t. Joseph Van Os/The Image Bank/Getty Images; b. Peter Weimann/Animals Animals/Earth Scenes. E12-E13: Craig J. Brown/Flashfocus. E16: l. NASA. E16: t. NASA/Earth Scenes. E18: PhotoDisc/Getty Images. E19: t. Art Resource. E26: t. Tony Freeman/PhotoEdit; b. John Coletti/Stock Boston. E27: c. Michael Groen for MMH. E30-E31: Dale Sanders/Masterfile. E31: B&C Gill Gillingham/Index Stock Imagery. E34-E35: Addison Geary/Stock Boston. E36-E37: Tom Salyer/Silver Image for MMH. E38: t.l. David Young-Wolff/PhotoEdit; b.r. John Eastcott/Yva Momatiuk/ DRK Photo. E39: David Matherly/Visuals Unlimited. E40: Dan Howell for MMH. E42-E43: Bob Daemmrich/Stock Boston. E46: t.r. Michael Newman/PhotoEdit; c.r. Tony Freeman/PhotoEdit; b.r. Siede Preis/ PhotoDisc/Getty Images. E46-E47: b. Eric Roth/Flashfocus. E47: t. Roger Wilmshurst/Frank Lane Picture Agency/CORBIS; c. CORBIS; b. Eric Roth/Flashfocus . E51: Washington Metropolitan Area Transit Authority. E52-E53: McCutchean/Visuals Unlimited. E54: Richard Hutchings/Photo Researchers, Inc. E55: Donald Specker/Animals Animals/Earth Scenes. E56: Mark Burnett/Stock Boston. E57: t., c. PhotoDisc/Getty Images; b. David Young-Wolff/PhotoEdit. E59: Jodi Jacobson. E62-E63: NASA. E64: b. John Neubauer/PhotoEdit. FO: SuperStock. F0-F1: Kunio Owaki/The Stock Market/CORBIS. F2-F3: Myrleen Ferguson/PhotoEdit. F6: c. C Squared Studios/PhotoDisc/Getty Images; b. PhotoDisc/Getty Images. F6-F7: RDF/Visuals Unlimited. F7: t.l. Spencer Grant/PhotoEdit; t.c. PhotoDisc/Getty Images; t.r. Diane Padys/FPG International/Getty Images; b. Wallace Garrison/Index Stock Imagery. F8: l. Spencer Grant/PhotoEdit; r. Diane Padys/FPG International/Getty Images. F10: l. VCG/FPG International/Getty Images; r. Stock Boston. F12-F13: Alan Kearney/FPG International/ Getty Images. F14: PhotoDisc/Getty Images. F15: c., r. PhotoDisc/Getty Images. F16: b. David Young-Wolff/PhotoEdit; r., inset Lawrence Migdale . F17: t.l., b.r. SuperStock; t.r. Hutchings Photography; b.l. Amanda Merullo/Stock Boston. F18: t. Dennis Gray/Cole Group/ PhotoDisc/Getty Images. F20: Peter Scoones/TCL/Masterfile. F21: l. Tony Freeman/PhotoEdit. F22: t.c.l. Steve Kline/Bruce Coleman, Inc.; dice Norman Owen Tomalin/Bruce Coleman, Inc. F22-F23: Bruce Byers/ FPG International/Getty Images. F23: t. Norman Owen Tomalin/Bruce Coleman, Inc. F24-F25: Spencer Grant/PhotoEdit. F26-F27: t. Bo Brannhage/Panoramic Images; b. PhotoDisc/Getty Images. F27: t. Burke/Triolo Productions/Foodpix; b. Spike Mafford/PhotoDisc/Getty Images. F28: t. D. Boone/CORBIS. F29: The Stock Market/CORBIS. F30: l. C Squared Studios; c. EyeWire; r. Siede Preis/PhotoDisc/Getty Images. F31: c.l. C Squared Studios/PhotoDisc/Getty Images. F32: t. Ernie Friedlander/Flashfocus/Index Stock Imagery; c.l. Gabriel Covian/The Image Bank/Getty Images; c. Fred J. Maroon/ Photo Researchers, Inc.; c.r. David Sieren/Visuals Unlimited; b.l. Stephen Shepherd/Alamy Images . F34-F35: t. Joel Sartore/Grant Heilman Photography; c. Leonard Lessin/Peter Arnold, Inc.; b. James L. Amos/Peter Arnold, Inc. F35: Gabe Palmer/The Stock Market/CORBIS. F37: l. Tom Pantages; r. Richard Megna/Fundamental Photographs. F38-F39: Glenn Vanstrum/Animals Animals/Earth Scenes. F40: Kim Fennema/Visuals Unlimited. F40-F41: Tom Bean/DRK Photo. F43: b. Bill Bachmann/PhotoEdit. F44-F45: b. Jack Hollingsworth/PhotoDisc/Getty Images. F46: t. Jerry Driendl/FPG International/Getty Images; c. SuperStock; b. Terje Rakke/The Image Bank/Getty Images. F47:

R 70